TECHNOLOGY

and society

TECHNOLOGY
and society

Jan L. Harrington
School of Computer Science
and Mathematics

Marist College

JONES AND BARTLETT PUBLISHERS
Sudbury, Massachusetts
BOSTON   TORONTO   LONDON   SINGAPORE

World Headquarters

Jones and Bartlett Publishers
40 Tall Pine Drive
Sudbury, MA 01776
978-443-5000
info@jbpub.com
www.jbpub.com

Jones and Bartlett Publishers
Canada
6339 Ormindale Way
Mississauga, Ontario L5V 1J2
Canada

Jones and Bartlett Publishers
International
Barb House, Barb Mews
London W6 7PA
United Kingdom

Jones and Bartlett's books and products are available through most bookstores and online booksellers. To contact Jones and Bartlett Publishers directly, call 800-832-0034, fax 978-443-8000, or visit our website www.jbpub.com.

Substantial discounts on bulk quantities of Jones and Bartlett's publications are available to corporations, professional associations, and other qualified organizations. For details and specific discount information, contact the special sales department at Jones and Bartlett via the above contact information or send an email to specialsales@jbpub.com.

This publication is designed to provide accurate and authoritative information in regard to the Subject Matter covered. It is sold with the understanding that the publisher is not engaged in rendering legal, accounting, or other professional service. If legal advice or other expert assistance is required, the service of a competent professional person should be sought.

Production Credits

Acquisitions Editor: Tim Anderson
Production Director: Amy Rose
Production Editor: Tracey Chapman
Editorial Assistant: Melissa Elmore
Senior Marketing Manager: Andrea DeFronzo
V.P., Manufacturing and Inventory Control:
  Therese Connell
Composition and Design: Spoke & Wheel

Illustrator: Accurate Artists, Inc.; Diana Coe
Cover Design: Diana Coe
Associate Photo Researcher and Photographer:
  Christine McKeen
Assistant Photo Researcher: Meghan Hayes
Cover Image: © gudron/ShutterStock, Inc.
Printing and Binding: Malloy, Inc.
Cover Printing: Malloy, Inc.

Library of Congress Cataloging-in-Publication Data
Harrington, Jan L.
  Technology and society / Jan L. Harrington.
    p. cm.
  Includes bibliographical references and index.
  ISBN-13: 978-0-7637-5094-7 (pbk.)
  ISBN-10: 0-7637-5094-8 (pbk.)
  1. Technology--Social aspects. I. Title.
  T14.5.H357 2008
  303.48'3--dc22
                                    2008019714
6048

Printed in the United States of America

12 11 10 09 08   10 9 8 7 6 5 4 3 2 1

Contents

CHAPTER 3 Technological Failures 55

CHAPTER 4 Resisting Technology 73

CHAPTER 5 Accessibility of Technology   85

CHAPTER 6 Economics and Work   103

CHAPTER 7   Human Behavior: Communicating and Interacting   139

CHAPTER 8 Government, Politics, and War   **173**

CHAPTER 9 Children, Education, and Libraries 197

CHAPTER 10 Science and Medicine   **213**

CHAPTER 11 Entertainment and the Arts   **243**

CHAPTER 12 Looking Ahead   **265**

Preface

Hello, and welcome to the book I've always wanted to write. Over the past 25 years I've written mainly technical books on everything from programming (introductory to advanced data structures), database management, data communications, and computer security, with a few Mac OS titles thrown into the mix. But this book is different. Although it's full of technology, it's not about how to use a specific technology but about how technology fits into the wider world.

This book grew out of a course I teach called Technology and Society. It's what we call the "capping course" for undergraduate information technology (IT) majors. Each major at the college where I teach has a capping course. It isn't really part of the major—it doesn't teach concepts in the major subject—but rather is the final course in a student's "core" curriculum, those distribution requirements that make a technology student take a couple of writing, a couple of history, a couple of social sciences...you get the drift. The capping course is intended to bring all those core courses together, along with the philosophy and ethics that the students have taken.

I've always looked at the capping course as an opportunity to get graduating seniors to turn their vision outward from the computer labs—where they're undoubtedly having a lot of fun—to look at the impact of what they do on the world around them. We talk about technology in a very broad sense, not just computers. We look at how technology has shaped human civilization and how civilization has shaped technology. The discussions pull knowledge the students have gained from ethics, philosophy, history, psychology, sociology, and, of course, writing and literature. The students use print, film, and online resources to research and write about a variety of current issues. They also conduct a research project with real human subjects that measures some use of current technology.[1]

This book brings together the wide range of topics we cover in the course. In it, you will find discussions of technological change, technologies that have failed and the impact of those failures, the distribution of technology users, as well as the impact of technology on a variety of aspects of our society (for example, the entertainment industry, economics, politics, education, and social interactions).

Although I've tried to be as even-handed as possible in the discussions, I do agree with the Neo-Luddites when they say that technology isn't neutral: There are both good and bad aspects to just about everything we humans create, manufacture, and synthesize. Which way the balance tips for any given technology depends on how we use it. Therefore, wherever possible, I've tried to include both sides of the story. Readers can then draw their own conclusions.

---

[1] Projects have included determining the proportion of students that are using technology (for example, cell phones and iPods) as they walk across campus, surveying unsecured wireless networks during a War Drive, and recording the severity and frequency of appearance of new viruses and other malware.

Just as the IT capping course draws from a wide range of subject areas, so does this book. It's full of references to history, psychology, sociology, anthropology, science, economics, philosophy, and ethics. It's been a joy to write; the research has been endlessly fascinating. I can only hope that you find the material half as intriguing as I do!

WHAT YOU CAN DO WITH THIS BOOK

What can you do with this book? (No, using it as a doorstop is not an option.) You can just read it because its contents are interesting. Or, if you're a teacher, you can use it as a text in a freshman course that introduces computing or as the basis of a course similar to our capping course. The issues raised in the book also make it appropriate to use in a computer ethics course or in a social science course that includes the impact of technology on society.

WHAT YOU CAN EXPECT FROM THIS BOOK

As a reader, you can expect this book to make you think about things you may have taken for granted. If you are a technology major—whether it is computer science, information systems or IT (or something else closely related), one of the sciences, or mathematics—it will get you to twist your mind into a new shape and let you look out from the computer or research lab to see how people who aren't technologically sophisticated react to the activities in which you intend to spend your working hours. If you are a social science major, it can help you view technology as an integral part of human society rather than something that has to be endured or ignored, if that is your persuasion.

This book assumes that you have at least some basic knowledge of technology. You should be comfortable with using a computer to do everyday tasks such as word processing and e-mail; you need to know what an operating system is and what it does. You should also be able to use the Internet, in particular, be familiar with using it to research a variety of topics.

As an instructor, you can expect this book to help you put your course together. Each chapter ends with "Things to Think About," discussion topics that are suitable for either written or oral assignments. Some require students to conduct simple live research; others may require students to do online research. The "Where to Go From Here" section, also at the end of each chapter, contains print and online sources that students can use to supplement what's in the chapter. In addition, the body of the text is full of footnotes with the URLs of articles that expand on discussions in the text (and include some interesting asides that I couldn't fit anywhere else but couldn't bear to leave out).

Other instructor's resources can be found on the book's Web site, including a description of the type of research project I use in my course and suggestions for further written and oral assignments. You'll also find live links to much of the video material referenced in the book.

A CAVEAT

This book contains literally hundreds of references to Web sites where students can find more information about topics covered in the text. All those links were alive when the book was sent to the publisher. However, the Web is a dynamic environment and there is simply no way to guarantee that every site will be active when someone tries to reach it. My apologies in advance for any broken links: They're the price we pay for the open, ever changing, World Wide Web.

ACKNOWLEDGMENTS

Writing this book has been an unabashed pleasure. The research as been fascinating and endlessly interesting. I therefore first want to thank Tim Anderson, my editor, and Jones & Bartlett, for giving me the opportunity to make this book a reality. And like most books, it wouldn't have happened without the help of Tim's excellent editorial assistant, Melissa Elmore.

The production staff has been great, including Tracey Chapman (Production Editor), Meghan Hayes (Photo Research), Pam Thomson (Copyeditor), and Charlotte Zuccarini (Proofreader).

While I was doing my research, many people passed me links and magazine and newspaper articles. Foremost among the eagle-eyed media watchers was Joy Rabin, who I would like to thank profusely for all her help.

Generating Change 1

WHERE WE'RE GOING

In this chapter we will

- Define technology.
- Discuss the accuracy of predictions of future technology made in the 1950s.
- Consider the difference between innovation that represents a true paradigm shift and innovation that is a refinement or reformatting of existing technology.
- Examine how difficult it is to generate a truly new idea.
- Discuss sources of innovation.
- Look at the role of science fiction literature in generating innovation.
- Consider how innovation is funded.

INTRODUCTION

Welcome to a book about technological change, both good change and bad change. There have been periods in human history when innovation was nearly absent—the so-called Dark Ages and much of the ancient Egyptian period.[1] Fortunately (or unfortunately, depending on your point of view), such periods haven't lasted. Since the dawn of the Renaissance, human society has entered a period of ever-accelerating scientific and technological advancement.[2]

We tend to think of technological change as beginning with the Industrial Revolution in Western Europe and North America. And, indeed, that is the period of change we will examine throughout this book, with most of our emphasis on 1950 forward. We will look at how those changes have changed us and the world we inhabit, as well as how humans have changed and constrained uses of technology. We will discuss where we are technologically and where we are likely to be in a few years time. (As you might expect, the further out in time you attempt to predict technology, the worse those predictions become.)

The purpose of this book is to help you see the effects of technology on our world. It will make it easier for you to think about the impact of the technology we use and to discover ethical and beneficial uses of that technology. In a world where sometimes technological change is so fast it's hard to keep up, it pays to stop for a little bit to examine where we've been, where we are, and where we're going.

JUST WHAT IS TECHNOLOGY?

Before we get started, let's take a moment or two to define what we mean by *technology*. Is it simply electronic devices, or is the term more encompassing? For the purposes of this book, technology is anything electronic but also anything created by technology. For example, this book considers pesticides created in a lab, genetically engineered organisms, many pharmaceuticals, and fabrics created from petroleum to be the products of technology and therefore fair game for discussion in this book. Technology's impact is therefore far wider than what is circumscribed by computers, video games, MP3 players, and so on.

Nonetheless, much of our discussion centers on the Internet. There are two reasons for this. First, no other technology has been so pervasive around the globe. Second, more research is done on the uses of the Internet than on any previous technology.

AND WHAT ABOUT FLYING CARS?

Technology does change fast but sometimes not as fast as we think it will. A popular type of television program in the 1950s was the "documentary" about technology of the future.[3] Shown in black and white—there was no color television at the time—such shows usually included kitchens with center consoles that contained dishes, cooking devices, food, and a dishwasher. The food came out and the dishes went in after the meal was eaten. The wife of the house wore her 1950s style dress, with a tidy apron tied at her waist, and was probably assisted by her grade-school-aged daughter.

---

[1] We tend to think of the ancient Egyptians as great inventors because they devised methods for creating the pyramids and other amazing monuments. However, ancient Egyptian society was very rigid and resistant to change. Once the ancient Egyptians figured out a satisfactory method of doing something, they were very unlikely to change, even if someone suggested something that was clearly more efficient.

[2] You will read an overview of technological development since the Renaissance in Chapter 2.

[3] For a print version of an article that predicted 50 years forward from the 1950s-and a good laugh—see http://blog.modernmechanix.com/2006/10/05/miracles-youll-see-in-the-next-fifty-years/. How funny is it? One of the predictions is that underwear will be recycled into candy.

In **Figure 1.1** you can see a typical "future" kitchen. (This particular image dates from 1956.) The domed structure at the front of the image is an oven from which the woman has just removed a completely baked and frosted birthday cake. Behind her, in the round structure, is the refrigerator and food storage unit. The kitchen also includes a machine that lets the cook insert a recipe card and then produces all the necessary ingredients.

It hasn't happened, at least not exactly in that way. The kitchen of 1950 didn't look significantly different from a modern kitchen. We can go to the grocery store and purchase prepared foods that we can then heat in the microwave, but most of us don't consider such foods as "home cooked" or healthy. They're fast convenience foods that we generally don't eat for every meal every day. That center kitchen console doesn't exist. In fact, the basic components of a kitchen haven't changed a great deal since 1950 (with the exception of the addition of microwave ovens). Socially, we *have* changed: Exactly who does the cooking and cleaning varies from one household to another.

Many predictions of future eating habits included food replaced with pills, just like "The Jetsons" television show. Even astronauts don't take pills instead of food. We humans like to eat; it's just that simple. We may have dehydrated foods for use in space, when camping, and by the military, but that usually isn't our meal of choice. We like hot human-made meals. This is one situation in which humans have said "no" to technological advances. In fact, we are actually willing to pay more for organic foods, which have been grown or raised without the use of the fertilizers and pesticides developed over the past 100 years.

The same future documentaries probably included a flying car. All our wheeled personal vehicles would be replaced with cars that used flyways rather than roadways. It hasn't happened—yet. We can't even manage to keep our cars from crashing into one another. Can you imagine the chaos if we were able to fly about and

Figure 1.1 A 1956 view of a future kitchen.

weren't constrained by roads? Roads are two dimensional; flyways are three dimensional! But NASA is working on an air traffic control system to manage such traffic, and the head of Moller, Inc., a company that produces personal flying vehicles such as that in **Figure 1.2**, believes there will be many of them by 2025.[4] Nonetheless, today's "flying cars" are intended to drive on roads like a car and then convert into airplanes when the driver/pilot reaches an airfield. They really aren't designed to replace cars used in commuting and trips to the grocery store.

And let's not forget about robots. In the future, every household would have a robotic servant to take care of all household chores. Robots could be teachers, police officers, assembly line workers…anything you can imagine. Although we don't have robotic servants, today's robots are closer to the dream of the 1950s than much of the other predicted technology. For example, the Scooba is a robot that washes floors (**Figure 1.3**). It doesn't look much like a person, but it's useful!

There are a number of major technological hurdles to creating the all-purpose humanoid home robot (the least of which is intelligence):

- Locomotion: Just walking up stairs is a tremendously complex task.
- Vision: Tracking a moving object whose background changes from light to dark is another exceedingly difficult problem to solve.
- Power: The hurdle is to provide power that is portable and that lasts a reasonable length of time.
- Regulating hand strength: How does a robot determine how hard to grip an object? A 20-pound rock needs a firmer grip than a child's hand, but telling the difference between the two without causing damage to the softer object is a challenge for a machine. Humans can look at an object, reference their stored knowledge about that object, and get instant feedback about how heavy the object might be and how strong a grip to use to lift that object. This is a difficult task for a robot.

The state-of-the-art in humanoid robots is ASIMO, a research project of the Honda Corporation (**Figure 1.4**).[5] With ASIMO, Honda first tackled the locomotion problem. ASIMO, which is about 3 feet tall, walks like a bent-kneed human. It can climb stairs in a relatively natural way and jog its way around traffic cones. Early versions of the robot were plugged in to external power, but now ASIMO wears its power on its back. Honda's goal is to make ASIMO the all-purpose human helper predicted in the 1950s, but you shouldn't expect an ASIMO in your kitchen next year. The robot of our dreams hasn't happened—yet.

Technology is changing faster than we can keep track of it. As new ideas and their implementations make their way into our world, the impact of those technologies often slips by us. Many of the changes are gradual: It often takes an entire generation for a new technology to be accepted as a normal part of our daily lives. Sometimes we resist the change for one reason or another (see Chapter 2). Our imaginations often outstrip what technology can do, but eventually, if we can imagine it, it can probably be done.

---

[4] You can find the entire article about NASA's personal air traffic control system at http://www.cbsnews.com/ stories/2005/04/15/60minutes/main688454.shtml

[5] For an update on ASIMO's development, see http://asimo.honda.com

Figure 1.2
A current personal
flying vehicle.

Figure 1.3 Scooba, a robot that washes
floors.

Figure 1.4 The picture of ASIMO walking designated as No. 23 by Honda.
Honda's ASIMO, a humanoid robot that
walks much like a human.

So, where are the flying cars? (At airports.) Where are the humanoid robots?
(Under development.) Why does technology change as fast as it does? Why doesn't
it move faster? What is technology doing to us as human beings? And what are
we doing to it? We look at all these questions throughout this book to help you
get a handle on how technology has and hasn't changed us.

PARADIGM SHIFTS AND OTHER EARTH-SHATTERING EVENTS

A *paradigm shift* is a major change in the way we think about or do something. For example, the invention of movable type caused a paradigm shift in the way in which books were made.[6] Rather than copying a book by hand or carving the images of all pages in an entire book out of wood, a printer could take a set of letters and arrange them into the words on a page and then rearrange the same set of letters for the next page. What a marvel! Books could be produced faster, in greater quantities, and at less cost.

The implications of this were staggering. Books that were too expensive for most individuals became more affordable, leading to a much wider dissemination of information. More affordable books meant that more people had the motivation to learn to read. Given that many historians and political scientists will tell you that a literate population is a prerequisite to democracy, we could say that the invention of movable type was the foundation of modern democratic governments.

Most of our technical change is incremental and doesn't represent a paradigm shift. Increasing the clock speed of a silicon-based microprocessor and increasing the number of transistors on the chip is an incremental change rather than a change in the fundamental underlying technology. In fact, when we look closely at technological change, we see that there have really been very few true paradigm shifts.

Exactly which ideas or inventions warrant being identified as paradigm shifts is certainly a matter of opinion, but most students of the history of technology would include the following:

- Movable type
- Boolean logic (the theory behind the use of 0s and 1s to represent and manipulate data)
- Babbage's difference and analytical engines (the first appearance of the architecture still used in modern computers)
- The cotton gin
- The transistor (the foundation upon which all of today's electronic devices rest)
- The graphic user interface (GUI)
- The Internet and the World Wide Web

There are some omissions from the preceding list that might seem strange to you, such as the inventions of Thomas Edison. Edison's inventions used existing technological ideas; he was the ultimate inventor but not the ultimate innovator. He didn't come up with the idea of the incandescent light bulb, for example, but he was the first to make it work.

Recognizing Paradigm Shifts

Given the pace of current technological change, we see many "inventions" or "innovations" every year. However, it isn't always clear whether a given new tech-

---

[6] The first verifiable movable type was made of clay in China, around 1041. However, it was Johannes Gutenberg's printing press of 1440, with its metal and wooden movable type, that jump-started Western printing.

nology represents a paradigm shift, is a refinement of an existing technology, or will have no lasting impact at all. Some technologies, especially those that are the result of concerted research efforts, are clearly important at the outset. The result of others can't be judged until time has passed to demonstrate their impact.

Consider, for example, the transistor. This seminal piece of technology was not an accident. Soon after commercial computers appeared, it was clear that to make them affordable and usable, something would be needed to replace the vacuum tube, which made the machines too big, too hot, and too expensive. Two commercial research labs developed the transistor almost simultaneously. There was no question in anyone's mind what the transistor was good for. The research efforts were designed to find something to replace vacuum tubes, and they did.

In contrast, the GUI wasn't recognized as a paradigm-shifting technology for several years after it became commercially viable. Xerox, which pioneered the mouse-driven point-and-click interface, was unable to produce an affordable computer. When the Macintosh debuted in 1984 with the first widely available GUI as part of its operating system, it was derided as "too cute." But slowly, over time, respect developed for the benefits of the GUI (as opposed to a command-line interface, such as that required by MS-DOS). Twenty-five years later, the GUI is the predominant user interface.

The impact of the Internet and the World Wide Web were similarly unpredictable. When we look back at predictions made about future technologies, virtually no one foresaw the Internet turning into the global network it is today, nor did the appearance of the first Web browser and Web server in 1990 have a major impact. It wasn't until the number of Web sites reached a point where ordinary users were finding the sites beneficial that the impact of the World Wide Web became clear.

We are faced with the same problem when it comes to evaluating most technologies that have appeared in the past few years. For example, consider the technology innovation awards presented by the *Wall Street Journal* each year. The 2007 awards went to the following:[7]

- Gold award: This award went to a drug for hypertension, developed by Tekturna, a Swiss company. The drug blocks the enzyme that triggers hypertension and can replace the multiple drugs taken by a patient.
- Silver award: The silver was awarded to a method for condensing drinking water from the air, developed by Aqua Sciences.
- Bronze award: This award went to a technology for providing high-resolution television viewing over the Internet (Joost, based in Luxembourg).

The honorable mention for computing was awarded to Ncomputing, which developed a way to share a personal computer (PC) with multiple users. Tellme Networks was also given an honorable mention for their cell phone technology that lets users request directory listings and other information with voice commands; the results appear as scrollable text on the mobile phone.

---

[7] For the complete article detailing the awards, see "The Journal Report" in the September 24, 2007 issue of the *Wall Street Journal*.

Which of these technologies constitutes a paradigm shift or will even have a lasting impact? It's too soon to tell. Certainly, none solves as massive a technological problem as that presented by vacuum tubes. Any easy way to extract water from air will be beneficial in arid regions, but will it produce a fundamental change in the way we operate? Very few of us are prescient enough to be certain we are making an accurate prediction. All we can do is wait to see what happens.

THINKING "OUTSIDE THE BOX"

So why have there been so few real paradigm shifts in recent history? It is because truly original thinking (thinking "outside the box") is really very, very hard.[8] Consider, for example, videotape cassettes. Do they represent a true paradigm shift or an evolutionary change? They are almost certainly the later. Videotape was invented in the United States and ran on large reel-to-reel machines. (The tape was one inch wide as opposed to the quarter-inch tape used in home videotape machines; the hardware itself was five feet wide.) The Japanese took the tape-recording technology and repackaged it into cassettes that, because of their ease of use and storage, quickly replaced reel-to-reel applications.

"New" operating systems aren't really new. In fact, we haven't had a paradigm shift in operating systems since the introduction of the GUI. And there hasn't been a true paradigm shift in central processing unit (CPU) architecture as of yet. The basic underlying internal structure of a microprocessor hasn't changed since the first one was invented by Intel, and PC CPUs conceptually aren't very different from mainframe CPUs.[9] They are still following an input–process–output logic, even if that logic is performed by multiple parts of the CPU; they're all still based on silicon.

Will this change? It has to. We've reached the limits of how many transistors we can put in a given area without overheating the machine using the CPU. And soon we will reach the limit of how many cores we put on a single chip. When that happens, silicon will have reached its limits, and we'll have to turn to a totally different technology. But what will it be? It's much harder for humans to think of something completely new than to extend and enhance existing ideas.

In 1936, H. G. Wells, known for being one of the most gifted prophets of the future in his time, authored a screenplay called *Things to Come*. Based on his book, *The Shape of Things to Come*, it is both a strong antiwar film and a representation of his vision of a utopia in the late 20th century.[10] While not a very good film, what makes it so interesting is which technological innovations he was able to predict and which he totally missed. The film correctly depicts flat-screen

---

[8] For a fascinating look at how people of the 1930s and 1940s viewed technology, browse through the Modern Mexchanix Web site at http://blog.modernmechanix.com/covers/.

[9] Mainframe CPUs have larger registers than microprocessors and therefore can usually handle much larger quantities than microprocessors. Until recently, mainframes also had more CPUs than PCs. However, with the introduction of multicore, multichip PCs, that distinction is disappearing. Mainframes, nonetheless, have more sophisticated I/O capabilities and are far better suited to handling large amounts of data than desktop machines.

[10] See http://blog.modernmechanix.com/2006/04/30/h-g-wells-things-to-come/
To view the film online, start at http://www.youtube.com/watch?v=z0pNV2gjhMQ

monitors and holographic projections. It also includes a helicopter (although it flies level; real helicopters fly with their noses down) and prefabricated building panels. But the film misses some of our most prevalent current technologies:

- Rocketry: The airplanes of Wells' future are propeller driven. The end of the film sends two people around the moon using a "space gun" that looks much like one of the big guns from a World War II battleship.
- Digital displays: Even projecting far into the future, gauges and clocks are all analog.
- Computers: There isn't a computing machine in sight.
- Weapons: The use of poisonous and benevolent gas continues. Wells did not foresee the global agreement not to use gas in warfare. In fact, his ultimate weapon is the "gas of peace," which puts the bad guys to sleep so the good guys can come in safely and stop the fighting. In addition, Wells certainly did not foresee nuclear weaponry.

Many of the things that Wells wasn't able to predict, such as rocketry, computers, and nuclear weapons, were developed within 15 years of the release of his film. Yet it took the needs of a war to make such paradigm shifts happen.

SOURCES OF INSPIRATION

There is an aphorism in the world of technology that says that innovation is "one-tenth inspiration and nine-tenths perspiration." To some extent, that would appear to be true: Once someone generates an idea or theory, it takes a lot of work to develop the idea or theory into functional technology. Often, like Edison, who tried hundreds of filaments for incandescent lights before hitting on a material that worked, it takes an enormously long time to turn an idea into a practical solution. But first, you need that one-tenth inspiration.

Does inspiration come just by sitting and thinking? Sometimes, but often something triggers a solution to a problem. Consider, for example, the problem facing Shell Oil when it came to extracting small oil deposits from under the sea in southeast Asia. It is too costly and environmentally unsound to develop oil rigs for every small deposit, yet the oil company didn't want to leave the oil in the ground.

The engineer assigned to work on the problem was Jaap Van Ballegooijen. Frustrated because he couldn't come up with a solution, he left southeast Asia and returned home to Amsterdam. He took his son out to lunch and watched as the teenager turned a bent straw upside down to suck up the dregs of a milkshake. The idea of a flexible tube that could suck up something became the inspiration for a new oil harvesting technology: a tube that could go under the sea.[11]

Other inspirations triggered by events include Isaac Newton seeing the fall of an apple and discovering gravity and Archimedes' observing how his own body displaced water in a bathtub, which led him to the discovery of how solids displace their own mass in water.

---

[11] Shell made a movie out of this story. You can view it at http://www.youtube.com/watch?v=AzJDDjA3AM4/ The film is, of course, a product of Shell public relations. Regardless of how you feel about big oil companies, the fact still remains that a new technology came from inspiration generated by a teenager's use of a straw.

At times researchers are looking for one thing and end up with another; they make discoveries or develop new technologies by accident. This is known as *serendipity*. One of the best known examples of serendipity involves the Phillips Petroleum Company (today known as ConocoPhillips) and the discovery of polypropylene, a compound that is now part of many plastics. Initially, Phillips wasn't in the plastics business at all. The company had assigned two research chemists— J. Paul Hogan and Robert L. Banks—to find better catalysts for refining high-octane gasoline using propylene and ethylene. On June 5, 1951 the two chemists discovered that their equipment was clogged with a white, tacky substance, which later became known as polypropylene. Fortunately for their employer, they knew what they had done and were able to record and replicate the procedure. Phillips obtained a patent on polypropylene (and polyethylene, which was developed shortly thereafter) and entered the plastics business.

Other items that had serendipitous beginnings include the following:[12]

- Cheese: An Arab carrying milk across a desert in a pouch made from a sheep's stomach discovered that during his journey the milk had turned into cheese.
- Popsicle: A boy left a soda on his front porch overnight and discovered the next morning that it had frozen into a delectable treat.
- Potato chips: A chef who was angry with a complaining customer kept cutting potatoes thinner and thinner for frying. Instead of complaining when the fried potatoes were razor thin, the customer loved them.
- Coke: Cola drinks (Coca Cola in particular) began as headache medicine. The doctor working on the project wanted to make something that tasted good as well as being effective. The medicine didn't work particularly well, but the taste was great. Carbonated water was added to turn it into a soft drink.
- Silly Putty: The developers were trying to find a rubber substitute for the United States during World War II.
- Teflon: The developer was trying to develop a new refrigeration gas but got the nonstick surface instead.
- Aspartame (now marketed as NutraSweet): The developer, James Schlatter, was attempting to create a test for an antiulcer drug. The result was a no-calorie sweetener.

ROLE OF SCIENCE FICTION

Some new technological ideas have come not from scientists and technology researchers (either at commercial labs or academic institutions) but from writers and, in particular, from science fiction writers. Both print and film have spurred technological innovation that we might not have considered otherwise.[13]

---

[12] For even more examples, see http://library.thinkquest.org/J0111766/accidents.html and http://en.wikipedia.org/wiki/Serendipity

[13] Most of us who were alive and aware of television in the 1960s saw at least some episodes of "The Jetsons," that far future Hanna-Barbera cartoon show. We also didn't take its view of the future seriously. It was little more than the same old 1950s predictions and 1960s social roles written as an animated situation comedy.

For example, the concept of geosynchronous satellites and the idea that it takes only three such satellites to provide coverage for the entire earth are the brainchildren of Arthur C. Clarke, a world-respected author of science fiction. It was the "Star Trek" television series that gave us the ideas for medical scanners and matter transportation ("beaming"). Our scanners aren't handheld—yet—but they do exist. And although the idea of disassembling an object or person, transmitting it digitally, and reassembling it at its destination is a dream, scientists have managed the feat with a single molecule.[14] We don't have transporters—yet.

Keep in mind, however, that not everything that comes from the mind of a science fiction writer or filmmaker is technologically or scientifically feasible. For example, we currently have many uses for lasers, but we can't make a "Star Wars"-style light saber. Why? Because the light saber requires a laser that travels a specific distance and then stops abruptly. But lasers don't work that way. The light keeps traveling until it either dissipates over distance or is stopped by something solid, such as a wall or a human body. As far as we know today, there is no way to stop the transmission of light before the light dissipates on its own without that physical barrier.

WHO'S PAYING THE BILLS?

It costs a lot of money to develop new technologies, taking them from theory and concepts to usable, marketable items. Who pays for it? There are three usual sources for funding: government, industry, and private firms and individuals. The proportion of technology research and development monies that come from each of those sources depends to a large extent on the country in which the research and development occur.

Government Funding

Many technological developments have come from government-funded projects. For example, the Manhattan Project, which developed the first atomic bomb, was totally funded by the U.S. government. In the opinion of those waging World War II for the United States, it was essential to bringing the war to a conclusion. In fact, the amount of money channeled to the Manhattan Project was so great that other segments of the U.S. economy were shortchanged.

As part of its World War II efforts, the British government funded cryptology research at Bletchley Park. The immediate purpose of the work was to find ways of deciphering German codes, but ultimately the work was fundamental to the development of computers over the next 20 years.

Perhaps the most important U.S.-funded project has been ARPANET, the project that ultimately led to the Internet.[15] The original project, funded by both the National Science Foundation (NSF) and Advanced Research Projects Agency (ARPA), was to develop a secure communications network for the U.S. military that could be used in an emergency, when other forms of communication were

---

[14] To find out where we are with beaming research, and the problems associated with actually transporting an entire human that way, see http://www.se51.net/devnull/2007/07/12/star-treks-beaming-becoming-reality/

[15] To be strictly accurate, the NSF was a major supporter of the ARPANET project. It had its own network, NSFNET, which was the network that evolved into the Internet.

unavailable. What was unusual about this project was that, for the most part, the government kept its hands off the project and allowed researchers to choose their own directions.

During the Cold War, almost all science and technology research performed in the Soviet Union was government funded. Not only was the Soviet economy unable to fund research outside the pubic sector, but government funding also gave the government control over the direction of the research.

Governments sometimes make technological achievement part of their agendas. The race to place a man on the moon, for example, was essentially a battle in the Cold War between the United States and the Soviet Union. U.S. President Kennedy issued a challenge to the country to put a man on the moon before 1970, before the Soviet Union could get there. NASA was given the funding it needed to reach that goal, and the United States successfully landed men on the moon in 1969.

In 1980 the Japanese government funded a 10-year project to develop artificial intelligence (AI), known as the Fifth Generation Project. Government scientists were certain that the goal was attainable and funded the project accordingly. True AI turns out to be far more elusive and complex a problem than anticipated; the Japanese project did not produce true AI within 10 years. However, it spurred artificial intelligence research in Japan, which continues today.

Government funding is often affected by politics. For example, consider the state of government funding for stem cell research in the United States. Stems cells, cells that can develop into many types of specialized human cells, have shown great promise in the treatment of diseases such as Parkinson's. However, the source of the best stem cells (those that can be most successfully coaxed into a desired form) is human embryos. The U.S. government has taken the position that it is unethical to use embryos for this purpose and has limited government funding for stem cell research to projects that use existing stem cell lines; the creation of new sources of stem cells or projects that use them will not be supported by federal funds.

Today, the U.S. government funds a number of technology research projects, many under the auspices of grants from the NSF. The projects for which funding is available are usually closely tied with government priorities, such as the use of technology in education, science and technology cooperation with Russia, data communications (in particular, security), and various projects for the development of military hardware. Some of the funded research is "basic" in that its aim is to advance theory; the rest is "applied," which is designed to produce actual technology products. Nonetheless, the government does not provide the majority of technology research funding in the United States: The largest proportion of the funding comes from industry.

Other countries provide a varying amount of technology and development research support. Sweden, for example, spends approximately 4% of its gross national product on research (one of the highest amounts in the world), but only one-quarter of that amount is provided by the government. In contrast, most research and development in Russia is funded by the government; few industries are economically strong enough to sustain their own research efforts. China, however, is just beginning to establish government agencies for funding research.

Industry

Much applied research and development in the United States comes from individual corporations that are looking for new product opportunities. Pharmaceutical companies, for example, devote a portion of their budget to developing and testing new drugs. In 2005 IBM spent approximately 5% of its revenue on research and development. In the same year, Microsoft spent newly 21% of its revenue on research and development.[16]

Some corporations do more than applied research, however: They engage in basic research through separate corporate divisions called corporate *think tanks*. These organizations don't necessarily produce marketable products but rather advance the science and technology of their fields. Three of the most well-known think tanks are as follows:

- Bell Labs:[17] Headquartered in Murray Hill, New Jersey, Bell Labs (once known as AT&T Bell Labs) was established in 1925 by AT&T and Western Electric. Researchers are credited with developing the radio telescope, the transistor, the laser, information theory, UNIX, the C programming language, and the first wireless network.
- IBM Watson Research Center:[18] IBM's think tank is headquartered in Yorktown Heights, New York. It began in 1945 with the goal of developing computing devices to help with the World War II war effort. Researchers are credited with developing carbon nanotube technology, a variety of microprocessors (including a collaboration with Motorola on the Power PC), the Data Encryption Standard (DES), the FORTRAN programming language, fractals, magnetic disk storage, and the theory of relational databases.
- Xerox PARC (Palo Alto Research Center):[19] Xerox began its PARC research center in 1970. Among the developments that have emerged from PARC's labs are the laser printer, the SmallTalk programming language, Ethernet, GUIs, and client/server computing,

Corporations fund think tanks for at least two reasons. First, the research often generates creative and innovative products. Second, the recognition earned by the researchers adds to the reputation of the corporation and ultimately translates into more sales. The research at Bell Labs, for example, has generated six Nobel prizes.

---

[16] You can find more research and development statistics at http://www.technologyreview.com/articlefiles/2005_rd_scorecard.pdf

[17] http://www.bell-labs.com (There is another organization named Bell Labs, but it is involved in controlling rodents.)

[18] http://www.watson.ibm.com/index.shtml

[19] http://www.parc.xerox.com/

Private Firms and Individuals

There are some private companies and wealthy individuals that support technology research and development. For example, Sir Richard Branson and Burt Rutan became the prime financial backers of a private effort to develop a spacecraft. Their SpaceShipOne completed two suborbital flights within a two-week period in 2004, earning a $10 million prize.[20] Their company (The Spaceship Company[21]) is now developing a fleet of such suborbital craft. As you might expect, there is far less private funding than that from governments or corporate sources.

---

[20] The X Prize was offered by the X Prize foundation, which is devoted to developing private space flight. See http://www.xprize.org/
[21] http://www.scaled.com/projects/tierone/

WHERE WE'VE BEEN

We are in the midst of a period of rapid technological change. Most of that change is incremental rather than representing a fundamental paradigm shift. It is extremely difficult to generate totally new ideas; we tend to refine existing technologies, making them faster, giving them more features, and, at the same time, making them cheaper. The predictions of technologies from 50 years ago have not come true, although in some cases—in particular, robotics—we are approaching what was predicted.

The ideas for change come from many places. Some comes from scientists working in laboratories, some from academic settings, and even from science fiction writers. The funding for technological development may come from government, corporate, or private sources.

THINGS TO THINK ABOUT

1. What do you consider to be a "technology"? Do you agree with the definition at the beginning of this chapter, or should the definition be more restrictive? Why?

2. Now it's your turn to attempt to think "outside the box." Spend a couple of hours trying to imagine a totally new technology. (It doesn't have to be computer related; it can be any kind of technology.) Then write up your ideas.

3. What are the latest gadgets in kitchen technology? (You can research this on the Internet.) Categorize each item you find as either a refinement of existing technology or a paradigm shift.

4. Imagine yourself a technology researcher. You have an idea for a totally new type of computer printer, but you don't have the money to develop it yourself. Use the Internet to find funding sources for your research efforts.

5. Read one of the following classic science fiction novels: *Foundation* by Isaac Asimov, *Dune* by Frank Herbert, *The Moon Is a Harsh Mistress* by Robert Heinlein, *Neuromancer* by William Gibson, *Ender's Game* by Orson Scott Card, or *Startide Rising* by David Brin. As you read, make notes about the future technology described in the book. When you have finished, comment on each technology, stating how likely you believe it is that we will actually see the technology.

6. Describe a technology that was developed "on purpose." In other words, research and identify a technology that was the focus of a concerted development effort, where what the researchers were attempting to create was actually created. Determine who funded the development and whether the technology has actually been brought to market.

WHERE TO GO FROM HERE

Facilitating Innovation and Creativity

Chesbrough, Henry William. *Open Innovation: The New Imperative for Creating and Profiting from Technology.* Cambridge, MA: Harvard Business School Press, 2003.

Christensen, Clayton M., Erik A. Roth, and Scott D. Anthony. *Seeing What's Next: Using Theories of Innovation to Predict Industry Change.* Cambridge, MA: Harvard Business School Press, 2004.

Davila, Tony, Marc J. Epstein, and Robert Shelton. *Making Innovation Work: How to Manage It, Measure It, and Profit from It.* Upper Saddle River, NJ: Wharton School Publishing, 2005.

De Bono, Edward. *Lateral Thinking: Creativity Step by Step.* New York, NY: Harper Paperbacks, 1973.

Ditkoff, Mitchell. "Back to the garden: An 8-step process for creating a culture of innovation." *http://www.innovationtools.com/Articles/EnterpriseDetails.asp?a=278*

Gelb, Michael J. *How to Think Like Leonardo da Vinci: Seven Steps to Genius Every Day.* New York, NY: Dell, 2000.

"Innovation: Life, Inspired" *http://www.pbs.org/wnet/innovation/*

Kelley, Thomas, and Jonathan Littman. *The Ten Faces of Innovation: IDEO's Strategies for Defeating the Devil's Advocate and Driving Creativity Throughout Your Organization.* New York, NY: Currency, 2005.

Michalko, Michael. *Cracking Creativity: The Secret of Creative Genius.* Berkeley, CA: Ten Speed Press, 2001.

Phillips, Jeffrey. "Creating a culture of innovation." *http://www.innovationtools.com/Articles/EnterpriseDetails.asp?a=276*

Rushkoff, Douglas. *Get Back in the Box: Innovation from the Inside Out.* New York, NY: Collins, 2005.

Yu, Larry, and MIT Sloan Management Review. "New insights into measuring the culture of innovation." *http://www.innovationtools.com/Articles/EnterpriseDetails.asp?a=277*

Funding Sources for Technological Advancement

The Australian Research Council, *http://www.arc.gov.au/*

"China to make research funding more transparent." *http://www.scidev.net/News/index.cfm?fuseaction=readNews&itemid=1600&language=1*

Funding a Revolution: Government Support for Computing Research. Washington, DC: The National Academies Press, 1999. *http://books.nap.edu/openbook.php?record_id=6323&page=36*

Lynn, Gary S., and Richard R. Reilly. "Growing the top line through innovation." *http://findarticles.com/p/articles/mi_m4070/is_2002_August-Sept/ai_91568346*

"The Swedish System of Research Funding." *http://www.vr.se/mainmenu/applyforgrantstheswedishsystemofresearchfunding.4.aad30e310abcb9735780007228.html*

Tassey, Gregory. "Comparisons of U.S. and Japanese R&D policies." *http://www.nist.gov/director/planning/r&dpolicies.pdf*

History of Technology 2

WHERE WE'RE GOING

In this chapter we will

- Consider the history of technology, beginning with humans' prehistoric efforts to make calculation easier.
- View important events in world history side by side with the history of technology to examine relationships between world events and the development of technology.

INTRODUCTION

Part of what you read about in this book concerns older technologies. To help you put those technologies in their historical context, this chapter contains a brief history of technology, along with some of the major world events that were happening at the same time. After a look at some early developments, a more detailed discussion follows, beginning in the 1800s, when industrial change began in Europe and the United States, and moving forward into the 21st century. Although it's difficult to know which recent technologies will be considered important down the road, we'll look at what the experts consider to be the seminal technologies of the past few years.

One of the things that may strike you as you read through this material is how early many inventions appeared. Things we assume are 20th century inventions—for example, the fax machine—were actually developed in the 19th century. However, in many cases we needed to wait for the widespread use of electricity to take full advantage of the inventions, and that didn't occur until the 20th century.

EARLY TECHNOLOGICAL EVENTS

Before 1800 there were a few important technological developments that acted as foundations for what came later. Without the inventions and discoveries listed in **Table 2.1**, much of our modern technology would not exist (or at least would have come much later).

Table 2.1 Early Technological Developments

Year	Major Events in the Development of Technology
Unknown, but prehistoric	• Early humans use bones as tally sticks. The earliest such bone found dates back to 35,000 B.C. (a baboon fibula with 29 notches).
~2400 B.C.	• The abacus is invented in Babylonia, the first known human counting device.
~500 B.C.	• First use of 0, apparently in India. • The first use of modern computing concepts such as recursion appears in India. The work of Panini, a grammarian, looks very similar to the grammars used today to specify the syntax of programming languages
~300 B.C.	• The binary number system is described by another Indian, the mathematician Pingala. His work includes the development of a binary code similar to Morse code.
~240 B.C.	• The Sieve of Erastosthenes lets people find prime numbers.
~200 B.C.	• The Chinese abacus (**Figure 2.1**), which is still in use today, appears.
~100 B.C.	• Negative numbers are first used, pioneered by Chinese mathematicians.
1045	• Movable type is invented in China.
1202	• Fibonacci, an Italian mathematician, introduces the Hindu–Arabic numbering system to Europe.
1206	• The first design for a robot appears (although the word robot came hundreds of years later).
1455	• A printing press that uses movable type is invented by Johannes Gutenberg.
1492	• Leonardo da Vinci diagrams aircraft. Argument still remains today over whether his designs would have flown.
1622	• The first slide rule is developed by William Oughtred. Based on a logarithmic scale, it simplifies multiplication and division. It also aids in logarithmic and exponential computations and can find square and cube roots (**Figure 2.2**).
1623	• Wilhelm Schickard devises the "calculating clock," a mechanism to add and subtract up to six-digit numbers. This was the first device that could detect overflow (when a number is too large to fit in the number of digits allowed).

Year	Major Events in the Development of Technology
1642	• French mathematician Blaise Pascal invents a mechanical adding machine (**Figure 2.3**). His intent was to help his father, a tax collector, with his work. Called the Pascaline, it could add and subtract only.
1671	• A calculating machine is developed by Gottried Wilhelm Leibniz.
1712	• A patent is issued for the first steam engine.
1745	• The first electrical capacitor (the Leyden jar) is invented.
1768	• A patent is issued for the spinning frame, an invention that will later become the focal point for antitechnology activity by the Luddites (see Chapter 4 for details).
1769	• James Watt improves on the steam engine, making it a practical device.
1774	• A patent is issued for the first telegraph.
1789	• The battery is developed by Alessandro Volta.

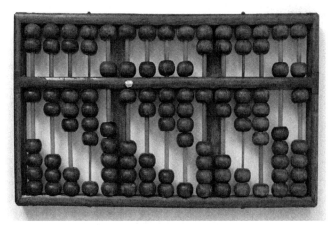

Figure 2.1
An abacus.

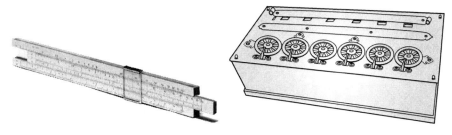

Figure 2.2 A slide rule, a mechanical device for multiplication, logarithms, and exponentiation.

Figure 2.3 A Pascaline, Pascal's adding machine (owned by the Musée des Arts et Métiers in Paris).

1800–1849

Many of the major technological advances and historical events of the first half of the 19th century can be found in **Table 2.2**. During this period many European empires fell as countries in South America became independent. Canada was consolidated into a single country, and the United States annexed a great deal of its western territory. European boundaries were in flux. The pace of technological change was relatively slow, although some fundamental items emerged, such as Jacquard's programmable knitting machines, Babbage's computing engines, passenger railroads, and the telegraph.

Table 2.2 Technology and Historical Developments During the First Half of the 19th Century

Year	Major Events in the Development of Technology	Major Events in World History
1800	• J. M. Jacquard invents a knitting machine—the Jacquard loom—that uses punch cards to tell the machine to change yarn colors. This ultimately leads to the use of punch cards for computer input. • Count Alessandro Volta's battery first appears.	• Napoleon takes control of France. • Washington, DC becomes the capital of the United States.
1803		• The United States completes the Louisiana purchase.
1804	• The first steam-powered locomotive, created by Englishman Richard Trevithick, appears but isn't practical because its extreme weight breaks the railroad tracks on which it is running.	• Haiti becomes an independent nation. • Napoleon turns France into an empire. • Alexander Hamilton is killed in a duel with Aaron Burr. • The Lewis and Clark expedition begins.
1805		• Lord Nelson defeats the Spanish Armada. • Napoleon continues to expand the French empire with victories over Austria and Russia.
1807	• Robert Fulton completes the first successful voyage of a steamboat.	
1808		• Napoleon conquers Italy and Spain.
1809	• The first electric light, an arc lamp developed by Humphry Davy, appears.	
1812		• Napoleon invades Russia and is defeated by the harsh Russian winter. • The War of 1812, between the United States and Great Britain, begins. The major issue is the right of U.S. vessels to travel unmolested in the world's oceans.

Year	Major Events in the Development of Technology	Major Events in World History
1814	• The first practical steam locomotive appears. • Joseph Nicéphore Niépce invents photography and uses his *camera obscura* to photograph the view out his window. It was very blurry—perhaps because the exposure required eight hours—but it was nonetheless what we consider to be a photograph. • Plastic surgery is performed for the first time (in England).	• Napoleon's defeat sees him exiled to Elba.
1815	• The stethoscope is developed.	• Napoleon returns for another try at being emperor. • After "100 days," Napoleon is defeated at Waterloo. • War of 1812 ends.
1819		• The area now known as Columbia, Venezuela, and Ecuador becomes independent of Spain.
1820		• The United States enacts the Missouri Compromise, which allows slavery in Missouri but bans it in the northern portion of the Louisiana Purchase.
1821		• Three Spanish colonies—Guatemala, Panama, and Santo Domingo—gain their independence. • Mexico declares its independence from Spain. It will become a republic in 1824.
1822	• Babbage designs his difference engine, a mechanical (but digital) device intended to solve polynomial functions. The engine was not built in the 1800s because the manufacturing of the day was not developed enough to fabricate the precise parts needed.	• Greece becomes a country separate from Turkey.
1823		• The United States publishes the Monroe Doctrine. Although its original purpose was to warn European nations not to interfere with the western hemisphere, it was used later to justify the annexation of territories by western hemisphere nations.
1824	• Balloons are invented as toys. • Portland cement is developed in England.	• Peru is liberated from Spain.
1825	• The first electromagnet is used. • In England, the first passenger railroad service begins.	

(continues)

Table 2.2 Technology and Historical Developments During the First Half
of the 19th Century (continued)

Year	Major Events in the Development of Technology	Major Events in World History
1826	• Photographs are taken on silver nitrate–based film. • The first U.S. railroad runs from Milton to Quincy, Massachusetts; the cars were pulled by horses.	
1827	• The first matches appear. • The microphone is invented by Charles Wheatsone.	
1829	• The typewriter is invented in the United States by W. A. Burt (**Figure 2.4**). • Braille printing is developed. • The first steam locomotive runs on rails in the United States.	
1830	• A Frenchman creates the first sewing machine.	• France invades Algeria and turns it into a French colony. • The French monarchy is abolished.
1831	• Canadian William Chalmers invents Plexiglas.	• Belgium becomes a nation separate from the Netherlands. • Nat Turner's rebellion against slavery in the United States is unsuccessful.
1832	• Louis Braille creates the stereoscope, a device for creating three-dimensional images.	
1833		• Great Britain abolishes slavery.
1834	• An ice-powered refrigerator appears. This first practical food preservation device uses ether. • Charles Babbage designs his "analytical engine," a device that was architecturally similar to computers of today. It includes a separate arithmetic logic unit, memory, and input–output devices. Like the difference engine, it could not be built using the technology of the day.	
1836		• Santa Ana and his Mexican army besiege the Alamo and kill all defenders. • Later in the same year, Texas becomes independent of Mexico. • South Africa becomes three nations (Natal, Transvaal, and Orange Free State).
1837	• This is similar to the telegraph key with which Samuel F. B Morse sent his famous May 24, 1844 message, "What hath God wrought!" Designed by his partner Alfred	• Victoria is crowned queen of Great Britain.

Year	Major Events in the Development of Technology	Major Events in World History
1837 (cont.)	Vail, the original was a simple strip of spring steel that could be pressed against a metal contact. Morse's use of this key changed the world of communication forever.	
1838	• Samuel Morse develops his Morse code for use over telegraph lines. Although the variable length code patterns make Morse code unsuitable for computer use, it is the first digital code in widespread use.	
1839	• Photography advances with the invention of daguerreotypes. • Kirkpatrick Macmillion creates the first bicycle. • The theory of hydrogen fuel cells is developed by Sir William Robert Groves (a Welshman).	• The Opium Wars between Great Britain and China begin. They will last until 1842.
1840		• Canada is united into a single nation.
1841	• The stapler makes it easier to keep papers together.	• Vice President John Tyler of the United States becomes president after President Harrison dies after only 1 month in office.
1842	• The development of the grain elevator eases the handling of farm products at ports. • Ether is first used as an anesthetic in surgery. • Canadian Benjamin Franklin Tibbets invents the compound steam engine.	
1843	• The first fax machine makes its debut, developed by Alexander Bain (a Scot).	
1844		• United States acquires the Oregon Territory.
1845	• American Elias Howe refines the sewing machine, making it practical to use.	• Texas is annexed by the United States.
1846	• Anesthesia is first used in dentistry.	• War begins between the United States and Mexico. The war will end in 1848 with Mexico giving up its claims to Texas, California, Arizona, New Mexico, Utah, and Nevada. • The United States annexes California and New Mexico.
1847	• Antiseptics are developed by Hungarian Ignaz Semmelweis.	
1848	• Boolean algebra, which is the basis for almost all computer operations today, was developed by British mathematician George Boole.	• Revolution in France sees Louis Napoleon elected as president. • The *Communist Manifesto* is published. • Harriet Tubman joins the U.S. underground railway and becomes a symbol for those looking to abolish slavery in the United States. • Upheaval occurs in Europe, with revolutions in Vienna, Venice, Berlin, Milan, Rome, and Warsaw.

Figure 2.4
An early typewriter,
circa the 19th century.

1850–1899

The second half of the 19th century (**Table 2.3**) is characterized by a more stable Europe. Borders of most of the countries we know today are set. The United States, however, must live through its Civil War and sort out the emancipation of slaves and its aftermath. In addition, industrialization brings worker unrest to factories, where the unions are beginning to take hold.

During this period technological developments occurred that form the underpinnings for much of our modern technology. There are advances in communications (the telephone and the wireless, the precursor to radio), entertainment (the phonograph and early moving pictures), and transportation (railroad crosses the United States).

Table 2.3 Technology and Historical Developments During the Second Half of the 19th Century

Year	Major Events in the Development of Technology	Major Events in World History
1850	• The dishwasher is invented, although it was not particularly practical at this point.	
1851	• Isaac Singer creates his first sewing machine. (**Figure 2.5**)	

Figure 2.5
Singer sewing machine.

Year	Major Events in the Development of Technology	Major Events in World History
1852	• The gyroscope is invented by Jean Bernard Léon Foucault.	• South Africa becomes a single nation.
1853	• George Cayley's manned glider flies successfully.	• The Crimean War begins when Turkey attacks Russia.
1854	• The principles of fiber optics are developed by John Tyndall. • The odometer is invented by Samuel McKeen in Canada.	• The Crimean War expands as Great Britain and France join Turkey against Russia.
1855	• A patent is issued to Isaac Singer for a sewing machine motor. • Rayon is invented.	• Violence erupts between slavery and antislavery supporters in Kansas, United States. • Florence Nightingale attends soldiers injured in the Crimean War.
1856	• Pasteurization is developed.	
1857	• Pullman sleeping cars, developed in Canada by Samuel Sharp, are first added to trains. • The undersea telegraph cable is invented by Canadian Fredrick Newton Gisborne.	• The U.S. Supreme Court issues the Dred Scott decision, stating that a slave is not a citizen. • The Sepoy Rebellion begins in India, resulting in rule by Great Britain.
1858	• First transatlantic telegraph cable is installed. • The rotary washing machine is invented.	
1859	• First internal combustion engine is created by Jean-Joseph-Étienne Lenior.	• Darwin publishes *The Origin of Species*. • In the United States, antislavery activist John Brown's attempted raid at Harpers Ferry fails. Antislavery rhetoric and activity is heating up throughout the country. • Italy is unified.
1860		• South Caroline becomes the first state to secede from the United States.
1861	• A patent is issued for elevator safety brakes to Elisha Otis. • The first modern bicycle appears. • Yale locks, developed by Linus Yale, appear.	• The Confederate States of America is formed by states that secede from the United States, triggering the start of the Civil War. • First U.S. income tax is enacted. • Russia frees its serfs.
1862	• The machine gun is developed by Dr. Richard Gattling. • The first plastic is developed.	• U.S. Civil War continues. • U.S. President Lincoln issues the Emancipation Proclamation, legally freeing the slaves.
1863		• U.S. Civil War continues. • French invade Mexico and capture the capital. The French install their own emperor.
1864		• U.S. Civil War continues.
1865	• The use of antiseptics in medicine is pioneered by Joseph Lister.	• U.S. Civil War ends. • U.S. President Lincoln is assassinated. • Mendel's *Law of Heredity* is published.

(continues)

Table 2.3 Technology and Historical Developments During the Second Half of the 19th Century (continued)

Year	Major Events in the Development of Technology	Major Events in World History
1866	• Dynamite is invented by Alfred Nobel. • A tin can with a key opener (like those found on sardine can lids) is developed.	• The Seven Weeks War occurs, resulting in the defeat of Austria by Prussia and Italy.
1867	• The first modern typewriter is developed by Christopher Scholes.	• United States buys Alaska. • Diamonds are discovered in South Africa. • Rule of the Shoguns ends in Japan. • The French pull out of Mexico; their chosen emperor (Maximillain) is executed. • Austria and Hungary are united under a single monarchy.
1868	• Air brakes are developed by George Westinghouse. • Traffic lights are invented by J. P. Knight.	• United States ratifies constitutional amendment giving civil rights to the freed slaves. • In Spain, Queen Isabella is deposed.
1869	• U.S. transcontinental train tracks completed.	• Suez Canal opens. • Mendeleev publishes his periodic table.
1870		• The Franco-Prussian War begins and continues until 1871. • France has yet again a new republic and a new government.
1871		• German Empire established. • Conflict with Native Americans in the United States escalates. • Graft is exposed in New York City; "Boss" Tweed arrested for fraud. • The Chicago fire kills 250 people. • African explorers Stanley and Livingstone meet.
1873	• Barbed wire is invented.	
1876	• Patent granted to Canadian Alexander Graham Bell for the telephone.	• George Custer is defeated by the Sioux at Little Big Horn.
1877	• Patented granted to Thomas Edison for the phonograph. The recording medium is a cylinder of tin foil. Each cylinder can be played only once. • Patented granted to Melville Bissell for a carpet sweeper.	• Rutherford B. Hayes becomes the first U.S. president to win the office by electoral college votes while losing the popular vote. • The Russian–Turkish war begins, ending in 1878.
1878	• First commercial telephone exchange opens in New Haven, Connecticut.	
1879	• Thomas Edison finds the right material for a light bulb filament and produces a practical incandescent bulb.	

Year	Major Events in the Development of Technology	Major Events in World History
1880	• First toilet paper appears. • John Milne invents the seismograph in Great Britain.	• A treaty between the United States and China allows the United States to limit Chinese immigration.
1881	• A metal detector is invented by Alexander Graham Bell. • Roll film for cameras appears. • A patent is issued for the automatic player piano.	• U.S. President Garfield is assassinated.
1882	• The tuberculosis germ is identified.	• Land evictions in Ireland result in terrorist activity. • Great Britain adds Egypt to its empire. • Standard Oil Trust becomes the first industrial monopoly.
1884	• The first fountain pen is developed by Lewis Edison Waterman. • The first cash register appears. Unlike modern cash registers, it is mechanical. • A patent is issued for the steam turbine.	
1885	• The first motor vehicle powered by an internal combustion engine is invented by Karl Benz. • The first gas-powered motorcycle is invented by Gottlieb Daimler.	
1886	• Herman Hollerith travels to France and sees Jacquard's punch card-controlled knitting machines. He realizes he can use the same concept to tabulate numbers. He devises a 12-position binary code that can be punched onto the cards (**Figure 2.6**) and also invents a tabulating machine to sort cards based on the punched values. • The first practical dishwasher appears. • The first practical four-wheeled motor vehicle is developed by Gottlieb Daimler. • Coca Cola appears.	• Anarchists bomb Haymarket Square in Chicago, killing seven. • The Statue of Liberty is dedicated in New York harbor.
1887	• The first radar appears. • The gramophone, which recorded on multiple-use wax cylinders, was invented by Emile Berliner. • The first contact lenses appear.	• Queen Victoria celebrates her golden jubilee.
1888	• George Eastman develops the first Kodak camera. • J.B. Dunlop develops the pneumatic tire. • The process for manufacturing drinking straws is developed. • The first alternating current motor and transformer are invented by Nikola Tesla.	• Jack the Ripper murders prostitutes in London.

(continues)

Table 2.3 Technology and Historical Developments During the Second Half
of the 19th Century (continued)

Year	Major Events in the Development of Technology	Major Events in World History
1889	• The matchbook is invented. • Smokeless gunpowder (Cordite) is invented.	• Eiffel Tower is built.
1890	• Hollerith's tabulating machine (**Figure 2.7**) is used by the U.S. government to analyze the 1890 census. • Patent issued for the diesel engine. • The first electric car heater is developed in Canada by Thomas Ahearn.	
1891	• The escalator is developed.	
1892	• The diesel-powered internal combustion engine is invented by Rudolf Diesel.	• Union activity escalates in the United States.
1893	• The zipper is invented.	• New Zealand grants women the right to vote (the first country in the world to do so).
1894	• Edison displays his kinetoscope (a precursor of movies).	• Sino-Japanese War begins (ends in 1895). • Frenchman Alfred Dreyfus is wrongly convicted of treason in a case that exposed underlying anti-Semitism. (He was pardoned in 1906.) • Union activity, including strikes in major industries, continues in the United States.
1895	• Wilhelm Roentgen discovers x-rays. • Moving pictures are demonstrated by the Lumière brothers.	
1896	• Patent issued in Great Britain for the first wireless, an early radio device (developed by Marconi). • Henry Ford constructs his first horseless carriage.	• Guilty over the use of the dynamite he invented, Alfred Nobel establishes the Nobel Prize for peace, science, and literature. • Athens, Greece hosts the first modern Olympic Games.
1897		• The Zionist movement begins.
1898	• The Curies discover radium and polonium. • A patent is issued for the first roller coaster.	• The Boxer Rebellion occurs in China. • The Spanish–American War begins (ends in 1899).
1899	• A patent is issued for the first motorized vacuum cleaner.	• The Boer War in South Africa begins (ends in 1902). • Union of South Africa is created as a part of the British Empire.

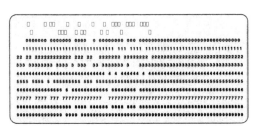

Figure 2.6 A typical punch card.

Figure 2.7 Hollerith's tabulating equipment.

1900–1949

The pace of technological change accelerated during the first half of the 20th century (**Table 2.4**). Researchers and inventors were able to build on the basic devices created during the preceding 50 years. In addition, the two world wars spurred the creation of many technologies, including computers (used for code cracking and navigation) and atomic weapons.

This was a period of some social change in the United States. Women received the right to vote, although their role as homemakers was largely unchanged. There was prosperity in the middle class, but their standard of living did not extend to those who were economically or socially disenfranchised. Although slavery had been eliminated, there was widespread discrimination against black people, who were by in large relegated to low-paying jobs, inferior public schools, and seats at the back of the bus. The neighborhoods in which people lived were typically segregated by race or ethnicity.

Table 2.4 Technology and Historical Developments During the First Half of the 20th Century

Year	Major Events in the Development of Technology	Major Events in World History
1900	• Count Ferdinand von Zeppelin invents lighter than air craft. • The modern escalator is invented.	
1901	• The first radio receiver receives a radio transmission.	• Queen Victoria dies. Her son, Edward VII, becomes king of Great Britain. • U.S. President McKinley is assassinated.
1902	• Air conditioning is invented by Willis Carrier. • Neon lights are invented in France by George Claude. • Enrico Caruso makes his first recording using a gramophone.	

(continues)

Table 2.4 Technology and Historical Developments During the First Half of the 20th Century (continued)

Year	Major Events in the Development of Technology	Major Events in World History
1903	• The Wright Brothers make the first powered airplane flight at Kitty Hawk, North Carolina. • Henry Ford establishes the Ford Motor Company.	
1904	• The tractor is invented by Benjamin Holt. • The first practical vacuum tube—the Fleming valve—is developed by John A. Fleming. • The New York subway opens.	• The Russo-Japanese War begins, triggered by conflicting claims to the ownership of Korea and Manchuria. The war ends in 1905. • Britain and France come to terms with each other.
1905		• Korea is returned to Japan and Manchuria to China. • The Russian Revolution begins.
1906	• American Lee de Forest invents the triode, the vacuum tube that will form the basis of early computers.	• An earthquake in San Francisco kills 500 and starts a fire that lasts for three days.
1907	• The Lumière brothers invent color photography. • The first helicopter to have a pilot appears.	• The rules of war are spelled out by the second Hague Peace Conference, in which 46 nations take part. • Oklahoma joins the United States as a state.
1908	• The first Model T Ford is sold. Ultimately, 15 million of the little black cars will travel on American roads.	• An earthquake in southern Italy and Sicily kills 150,000. • The U.S. Supreme Court clamps down on unions by ruling against secondary union boycotts.
1909	• Geoffrey W. A. Drummer publishes a design for an integrated circuit. Even as late as 1956 he is unable to build the device.	• Robert Peary and Matthew Henson reach the North Pole. • The NAACP is founded.
1910	• Thomas Edison shows a prototype of talking motion pictures.	• The Boy Scouts of America becomes an official organization.
1911	• Electric ignitions appear for automobiles. • Aircraft are first used as weapons in a war (the Turkish-Italian War). • Ernest Rutherford's research into the structure of the atom is successful.	• Italy annexes Libya after defeating the Turks. • The Chinese monarchy is overthrown, and Sun Yat-sen becomes president of the Chinese Republic. • The Mexican Revolution occurs. • A fire at the Triangle Shirtwaist Company in New York kills 146 factory workers. • Roald Amundsen finishes his journey to the South Pole.
1912	• Motorized movie cameras are developed to replace hand-cranked models. • The first military tank appears.	• The steamship Titanic sinks on her maiden voyage. • War in the Balkans ends with European Turkey being partitioned among Bulgaria, Serbia, Greece, and Montenegro.

Year	Major Events in the Development of Technology	Major Events in World History
1912 (cont.)		• Bulgaria invades Serbia and Greece but is defeated when Romania joins the coalition of defenders. • New Mexico and Arizona become U.S. states.
1913		• A garment workers strike in New York City ends with better conditions for the workers.
1914	• The gas mask is invented. • The first moving assembly line is put into service in the Ford assembly plant.	• World War I begins, triggered by the assassination of the Austrian Archduke and his wife. The initial conflict pits Austria against Serbia, Germany against Russia and France, and Britain against Germany. • The United States intervenes in a Mexican civil war, citing the need to protect American interests.
1915	• Einstein publishes his general theory of relativity.	• World War I continues. • A German submarine sinks the passenger ship *Lusitania*. • Turkish soldiers commit genocide, killing around 600,000 Armenians.
1916	• Radio tuners, which allow radios to receive multiple stations, are invented. • Henry Brearly invents stainless steel.	• World War I continues. • First birth control clinic is opened by Margaret Sanger. • The Easter Rebellion takes place in Ireland; British troops prevail.
1917		• The United States enters World War I, declaring war on Germany. Later in the year, the United States also declares war on Austria-Hungary. • The Russian Revolution takes place, marking the end of the Russian monarchy. Later in the year, the Bolsheviks stage a successful coup, with Lenin and Trotsky in control. The Bolsheviks then negotiate a peace with Germany. • A Jewish homeland in Palestine is promised by the Balfour Declaration.
1918		• World War I fighting ends. • German Kaiser abdicates.
1919	• The design of the flip-flop circuit appears. This circuit is used extensively in modern computers. • Pop-up toasters appear. • First nonstop flight across the Atlantic occurs.	• The Soviets establish control over worldwide Communist movements. • The Versailles Treaty is signed by the allies and Germany. • Gandhi begins his nonviolent crusade against British rule in India.
1920	• John T. Thompson invents the Tommy gun.	• The League of Nations begins meeting. • The U.S. states ratify the constitutional amendment giving women the right to vote.

(continues)

Table 2.4 Technology and Historical Developments During the First Half
of the 20th Century (continued)

Year	Major Events in the Development of Technology	Major Events in World History
1921	• The first robot is built.[1]	• The United States executes convicted spies Nicola Sacco and Bartolomeo Vanzettti.
1922	• People view the first three-dimensional movie.	• Mussolini takes control of Italy and establishes a Fascist government. • The Republic of Ireland (originally known as the Irish Free State) becomes its own nation. • The last Turkish sultan is overthrown by Kemal Atatiirk.
1923	• The traffic signal is invented by Garrett A. Morgan. • Vladimir Kosma Zworykin invents the cathode ray tube, which he calls an "iconoscope."	• French and Belgian troops occupy part of Germany to ensure that Germany makes its World War I reparations payments.
1924	• The first logical AND gate appears. AND gates (and other electronic gates based on Boolean operations) are used extensively in modern computers.	• Lenin dies, paving the way for Stalin to take control of the Soviet Union.
1925	• A Scottish inventor, John Logie Baird, demonstrates that human features can be sent to a television. • Canadian Arthur Sicard invents the snowblower. • The zipper is invented in Canada by Gideon Sundback.	• John T. Scopes is convicted of teaching evolution. • Adolf Hitler publishes the first part of *Mein Kampf*.
1927	• Canadian Frederick George Creed develops the first combined sending and receiving teleprinter. • An electronic television system appears, the precursor to our current television systems. • Technicolor (principally for film at this time) appears. • Lindbergh completes the first solo transatlantic flight. • The Big Bang theory of the origin of the universe is proposed by George Lemaître.	• The Russian Communist party expels Trotsky. • The collapse of the German economy paves the way for the Nazi Party's rise to power.
1928	• Penicillin is discovered by Alexander Fleming. • Canadian Morse Robb patents the electric organ.	
1929	• The theory of an ever-expanding universe is proposed by Edwin Powell Hubble.	• The U.S. stock market collapses, just one sign of major economic problems throughout the world.
1930	• Vannevar Bush of MIT develops an analog computer. This machine is a dead end; today's computers are overwhelmingly digital.	• Nazi power grows in Germany.

---

[1]The word "robot" was introduced in a play (R.U.R., Rossum's Universal Robots) by Karel Capek. It was first performed in 1921.

Year	Major Events in the Development of Technology	Major Events in World History
1930 (cont.)	• The jet engine is developed independently by Frank Whittle and Hans von Ohain. • Ernest O. Lawrence develops the first cyclotron.	• Nazi power grows in Germany.
1931	• The electron microscope is invented in Germany. • Heavy hydrogen is discovered by Harold C. Urey.	• Japanese occupy Manchuria. • Spanish monarchy is overthrown, allowing Spain to become a republic.
1932	• Edwin Herbert Land develops Polaroid photography. • The radio telescope appears.	• Nazis continue to gain power in Germany. • The Soviet Union suffers from famine.
1933		• Hitler becomes chancellor of Germany. • Nazis begin purging those considered undesirable. • League of Nations begins to fall apart as Germany and Japan withdraw. • The United States recognizes the Soviet Union.
1934	• The tape recorder (the first magnetic recording device) is invented.	• Nazis assassinate the Austrian chancellor.
1935	• A patent is issued for radar.	• Ethiopia is invaded by Italy. • The Nazis pull out of the Versailles Treaty that concluded World War I.
1936	• Alan Turing proposes the Turing machine, a theoretical concept that can be used to represent the logic of computing. A Turing machine manipulates symbols that are stored externally. The rules for symbol manipulation are expressed in formal logic notation. • Voice recognition is developed by Bell Laboratories.	• Germany moves into the Rhineland. • Italy annexes Ethiopia. • The Spanish Civil War begins, ending in 1939 with Franco's victory. • War begins between China and Japan. • Germany, Japan, and Italy become allies.
1937	• George Stibitz develops a 1-bit adder, a logical design to add two binary digits. • The first photocopier appears. • The first jet engine is designed and built.	• German military power continues to rise. • Germany unites Austria and Germany.
1938	• The first mechanical, programmable computer to use binary is developed by German Konrad Zuse. Originally named the V1, it could handle floating point numbers (up to 16 at a time). Input was read from holes punched in movie film or typed on a numeric keyboard; output appeared in lights. This machine was later known as the Z1.	• Germany invades Austria. • Czechoslovakia falls and Britain, France, and Italy allow Germany to partition the country.

(continues)

Table 2.4 Technology and Historical Developments During the First Half
of the 20th Century (continued)

Year	Major Events in the Development of Technology	Major Events in World History
1939	• John Vincent Atanasoff and Clifford Berry develop a 16-bit adder (digital circuits that can add two numbers of up to 16 bits each). The prototype is the first known machine to use vacuum tubes. This machine is considered by many to be the first electronic computing device, although it was not a general-purpose, programmable device. • Zuse completes his second computer, the Z2. Like the Z1, it can't repeat actions (loops) and therefore isn't a general-purpose Turing machine. The Z2 was later destroyed by Allied bombing during World War II. • Igor Sikorsky demonstrates his working helicopter. • Einstein talks to the U.S. president about the concept of an atomic bomb.	• Germany invades Poland. • Germany signs nonaggression agreement with the Soviet Union. • Germany voids agreements with Great Britain. • World War II begins. United States remains neutral.
1940	• Color television is invented. • NBC broadcasts the first network television shows.	• Germany invades Norway, Denmark, the Netherlands, Belgium, Luxembourg, and France. • Russia annexes Estonia, Latvia, and Lithuania. • The United States initiates a military draft.
1941	• Zuse develops the Z3 computer, which is programmable. It could repeat actions unconditionally but could not do operations such as "repeat until true." Given its architectural similarity to today's computers, the Z3 competes with the Atanasoff/Berry machine for the title of "first computer." • Enrico Fermi develops the first nuclear reactor (the "neutronic reactor") as the natural outgrowth of his research into nuclear fission.	• Japanese attack Pearl Harbor, bringing the United States into the war. Great Britain also declares war on Japan. • Germany invades Russia and the Balkans. • The Manhattan Project is established to develop an atomic bomb.
1942	• Atanasoff and Berry build a calculator that solves linear equations. Its significance lies in its architecture, which includes drum memory for main memory and punch cards for secondary storage. • Fermi produces a sustained nuclear chain reaction.	• The Nazi Holocaust is in full force. • United States rounds up Japanese and those of Japanese descent and moves them to internment camps.
1943	• The British code breakers at Bletchley Park create the Heath Robinson, a counting device used to break codes. It is the foundation for later code-breaking devices, especially Colossus.	• Mussolini is removed from office.

Year	Major Events in the Development of Technology	Major Events in World History
1944	• The Harvard Mark I is built by IBM. Also called the Automatic Sequence Controlled Calculator (a portion of which can be seen in **Figure 2.8**), it is the brainchild of Howard Aiken. The second programmable computer, it is based on vacuum tubes and mechanical relays. It uses punch cards, paper tape, and typewriters for input and output. Grace Hopper, a U.S. Navy officer, writes the world's first assembler for this machine. • Colossus, the code-breaking machine, enters use at Bletchley Park. Because of security surrounding the war effort, little was known about it until recently. • The kidney dialysis machine appears.	• The Normandy invasion takes place on June 6. • A proposal for the United Nations comes from the United States, Great Britain, and the Soviet Union.
1945	• Konrad Zuse develops the first high-level programming language (Plankalkül, or "Plan Calculus"). • Vannevar Bush develops the first hypertext system (his "memex"). • John von Neumann publishes a paper about the design of a computer whose architecture becomes the prototype for modern computers (the "von Neumann architecture"). • The design of the atomic bomb is completed. • Canadian Hugh Le Caine develops the first voltage controlled music synthesizer.	• Roosevelt, Churchill, and Stalin meet at Yalta to coordinate plans for the final defeat of Germany. • Hitler commits suicide. • Germany surrenders. • The first atomic bombs are dropped on the Japanese cities of Hiroshima and Nagasaki. • Japan surrenders. • The United Nations is established.
1946	• John W. Mauchly and J. Presper Eckert, working at the University of Pennsylvania's Moore School of Engineering, complete ENIAC (**Figure 2.9**). It is totally electronic but works in decimal numbers and has to be reprogrammed with patch cords for each program. Its 18,000 tubes, 7,200 crystal diodes, 1,500 relays, and 70,000 resisters weighs 30 tons and occupies 1,800 square feet. The vacuum tubes are constantly burning out because of the heat of the unit.[2] • Percy Spencer invents the microwave oven.	• The League of Nations is dissolved as the United Nations meets for the first time. • The Italian monarchy is abolished. • The Nuremberg war crimes trial concludes with guilty verdicts.

(continues)

---

[2]For a long time, ENIAC was considered to be the first computer. This was somewhat political, given the relative positions of Germany and the United States after World War II: The United States wasn't eager to give precedence to the German Zuse.

Table 2.4 Technology and Historical Developments During the First Half
of the 20th Century (continued)

Year	Major Events in the Development of Technology	Major Events in World History
1947	• Working at AT&T's Bell Laboratories, William B. Shockley, John Bardeen, and Walter Brattain invent the transistor. • The Mark II is completed at Harvard.[3] Grace Hopper, a U.S. Navy officer working on the project, creates the first assembler. • AT&T develops the idea of cellular phones. • Chuck Yeager flies faster than the speed of sound.	• India and Pakistan becomes independent of Great Britain. • The United States enacts the Truman Doctrine to help Greece and Turkey against Communist expansion. • The United States enacts the Marshall Plan to help rebuild Europe.
1948	• Small-scale experimental machine, or SSEM, developed at the University of Manchester, becomes the first machine to store both its program and data in RAM. • IBM introduces its "604" machine, which is the first computer to be built from parts that can be swapped on location.	• Burma and Ceylon become independent of Great Britain. • The Organization of American States is established. • The United Nations creates Israel and the Republic of Korea. • Gandhi is assassinated. • The blockade of Berlin by Communist forces begins; the West supplies the city by airlift. • Indonesia comes into being. • Soviet Union takes over Czechoslovakia.
1949		• NATO is established. • West Germany comes into being. • The Soviet Union tests its first atomic bomb. • Communist People's Republic of China comes into being. • The Soviet Union establishes East Germany.

Figure 2.8 A portion of the Harvard Mark I.

Figure 2.9 ENIAC.

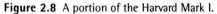

[3]The Mark II is the source of a story that deals with the first use of "bug" to describe a problem with a computer. Grace Hopper's assistant went to investigate a burned out vacuum tube and found a dead moth. He pressed the insect into the machine's log and wrote: "First actual case of bug being found." Is this true? We do have a photo of the log page (**Figure 2.10**), but even if it isn't, it sure makes a good story!

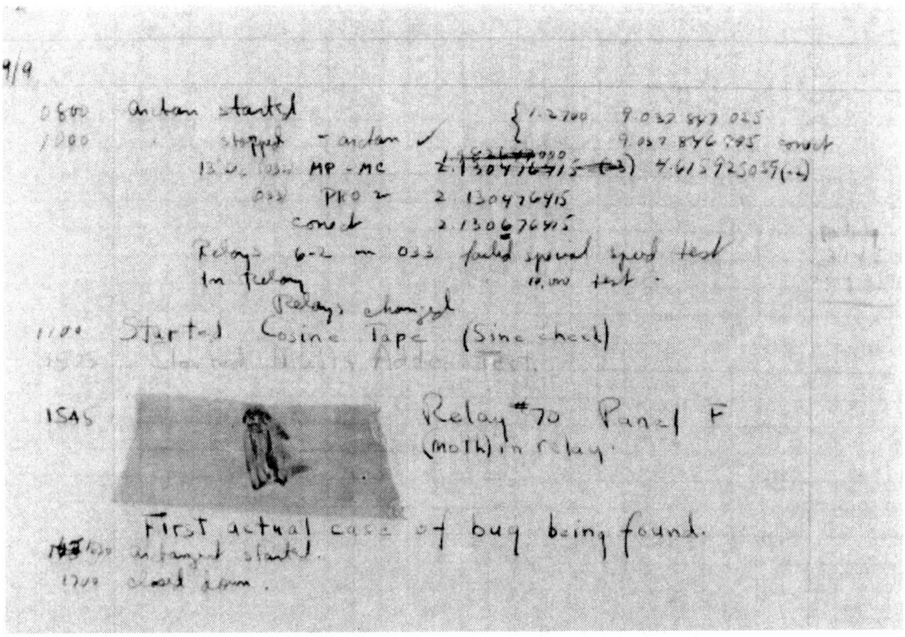

Figure 2.10 A log page from the Mark II, showing the first computer "bug."

1950–1999

When Americans over the age of 40 think back to the 1950s, they often see an idyllic, calm period. It was true that if you where a white, middle-class, American male, with a stay-at-home wife, you probably lived that type of life. However, if you didn't fit the model, then your life was more of a struggle, both socially and economically.

The 1960s heralded a period of rapid social change. Women broke out of their traditional roles, minorities became visible in their fight for an end to segregation, and the "Hippies" protested social and political inequity (especially the U.S. involvement in Vietnam). The momentum for these changes carried over into the 1970s. The last two decades of the century saw a world adjusting to the social upheaval of the preceding 30 years. It also saw the failure of Communist economic systems and significant political changes as Communist economies crumbled.[4]

The social changes were accompanied by rapidly accelerating technological developments. As shown in **Table 2.5**, the latter half of the 20th century saw the rise of modern computing, with quantum leaps in processor and storage power. Mainframes were challenged by desktop machines, the Internet came to dominate data communications, and we were flooded with the vast amount of information available on the Web.

---

[4]There are very few Communist governments left in the world today. China, the largest, has abandoned the socialist economic system in favor of a free market economy and is doing well economically because of that change. The other major Communist government is Cuba, which—depending on who you talk to—is either a paradise or a country where only the leader's cronies have a high standard of living. As with many such things, the truth is probably somewhere in between.

Table 2.5 Technology and Historical Developments During the Second Half of the 20th Century

Year	Major Events in the Development of Technology	Major Events in World History
1950	• Development of the hydrogen bomb begins. • Punch cards and their associated punch card machines (**Figure 2.11**) are the most common type of computer input device.	• Korean War begins.
1951	• UNIVAC (i.e., universal automatic computer) goes to work analyzing the U.S. census. Made by Remington Rand, it was the first "mass-produced" computer (**Figure 2.12**). • The British machine LEO (Lyons Electronic Office) becomes the first computer installed in a business. • The videotape recorder is invented by Charles Ginsburg.	• Libya becomes an independent nation. • The United States sentences accused spies Julius and Ethel Rosenberg to death. They are executed in 1953, although it is proven later that they were innocent. • Forty-nine nations sign a peace treaty with Japan.
1952	• IBM introduces the first machine in its 700/7000 series, the IBM 701. This is generally regarded as the first IBM mainframe. • A patent is issued to Joseph Woodland and Bernard Silver for the bar code. • The first hydrogen bomb is completed. Atomic tests take place on Enewetak.	• Elizabeth II becomes queen of Great Britain.
1953	• The first music synthesizer appears. • Flight recorders (the black box) are invented. • Texas Instruments invents the transistor radio. • The double helix structure of DNA is reported by researchers Watson and Crick. • United States tests a hydrogen bomb. • The Salk vaccine for polio becomes available.	• Korean War ends. • An uprising against Communist rule in East Germany is put down by government forces, using tanks. • Egypt becomes a republic.

Figure 2.11 An IBM punch card machine.

Figure 2.12 UNIVAC.

Year	Major Events in the Development of Technology	Major Events in World History
1954	• IBM introduces the IBM 704, which used magnetic core memory. • The solar cell is invented. • The United States launches its first nuclear submarine.	• The Soviet Union allows East Germany to exist as a separate nation. • The McCarthy hearings take place in the United States. • U.S. Supreme Court bans racial segregation in public schools. • SEATO is established. • Algeria begins to fight for its independence from France.
1955	• Maurice Wilkes invents microprogramming, which can be used to code the instruction set of a computer. • Optical fiber is developed.	• Rosa Parks refuses to give up her seat in the front of a bus, triggering one of the first organized protests against segregation. • Perón is thrown out of Argentina. • West Germany becomes a nation.
1956	• IBM introduces the first magnetic disk (RAMAC, for random access method of accounting and control). • George Devol and Joseph Engelberger form the first robotics company. • Christopher Cockerell invents the hovercraft. • A hydrogen bomb is tested in the atmosphere.	• Polish workers protesting Communism are stopped by government forces. • Egypt takes control of the Suez Canal. • A U.S. brokered cease fire stops the British, French, and Israelis from invading Egypt. • Morocco becomes an independent nation.
1957	• The IBM 704 becomes the first computer to use a high-level general purpose programming language (FORTRAN). • The Russians are the first to successfully place a spacecraft in orbit (*Sputnik I*). • Ken Olsen and Harlan Anderson found Digital Equipment Corporation (DEC).	• Nine black students in Little Rock, Arkansas enroll in a previously all-white high school.
1958	• The design of the modem is completed and modems are put into use by the U.S. military. • The laser is invented by Gordon Gould. • The first video game—Tennis for Two—is played on an oscilloscope at Brookhaven National Laboratory. • The United States launches its first satellite (*Explorer I*). • Canadian Joseph-Armand Bombardier invents the snowmobile.	• The European Common Market is established. • Egypt and Syria become the United Arab Republic. • Charles de Gaulle becomes premier of France.
1959	• Jack Kilby, working for Texas Instruments, applies for a patent for an integrated circuit. Robert Noyce, working for Fairchild Semiconductor, develops the integrated circuit simultaneously but does not apply for a patent until 1961.	• Fidel Castro assumes power in Cuba. • The Dalai Lama flees to India from Tibet. • Alaska and Hawaii achieve statehood.

(continues)

Table 2.5 Technology and Historical Developments During the Second Half of the 20th Century (continued)

Year	Major Events in the Development of Technology	Major Events in World History
1960		• A U.S. spy plane is shot down over Russia. The pilot is freed in 1962 in a prisoner exchange for a Soviet spy. • China and the Soviet Union go their separate ways because they interpret Communism differently. • African nations (Senegal, Ghana, Nigeria, Madagascar, and Zaire) become independent. • United States sends "military advisors" to Vietnam.
1961	• The first industrial robot is put into use at a General Motors automobile plant in New Jersey. • The Soviet Union successfully sends the first man into orbit. • United States launches two manned suborbital spacecraft.	• United States invades Cuba (Bay of Pigs) but is repulsed by Cuban forces. • The Berlin Wall is built between East and West Berlin by the East Germans to keep people from fleeing to West Germany. • More U.S. "military advisors" go to Vietnam.
1962	• The modem (the Bell 103, which had a speed of 300 bps) is sold to civilian users. • The audio cassette first appears. • The first computer game appears (Spacewar) and runs on a Digital PDP-1.[5] • First American spacecraft orbits the Earth.	• Algeria becomes an independent republic. • First black student enrolls at the University of Mississippi. • More nations gain independence (Burundi, Jamaica, Western Samoa, Uganda, and Trinidad and Tobago).
1963	• The first robotic arm to be controlled by a computer appears. • Bell Punch Co. Ltd. and Sumlock-Comptometer Ltd. of England release the "Anita," the first fully electronic calculator. • Videodiscs are invented. • The first artificial heart is placed in a human. The device is connected to machinery outside the patient's body.	• U.S. Supreme Court outlaws prayer in public schools. • Martin Luther King, Jr., delivers his "I have a dream" speech in Washington, DC. • U.S. President Kennedy is assassinated. • Kenya becomes independent.
1964	• The BASIC programming language is developed at Dartmouth College by John George Kemeny and Tom Kurtz. • DEC introduces the first successful minicomputer, the PDP-8. • Sony releases the first all-transistor calculator. • ANSI (American Standards Institute) makes ASCII the standard code for characters in a computer.[6]	• Three civil rights workers are murdered in Mississippi; seven of the 21 arrested are convicted. • Nelson Mandella begins a life sentence in a South African prison. • The Beatles come to the United States.

[5]There is only one working PDP-1 known to exist, but you can get a feel for this first game by playing it online at http://www.cgl.uwaterloo.ca/~anicolao/sw/Spacewar/spacewar.html
[6]The biggest alternative to ASCII is EBCDIC, which is used solely by IBM mainframes.

Year	Major Events in the Development of Technology	Major Events in World History
1964 (cont.)	• IBM introduces the 360 mainframes (**Figure 2.13**). This line is the first time that computers have belonged to a "family," where software is upwardly compatible as machines are enhanced.	
1965	• James Russell invents the compact disc. • Eight U.S. states and two Canadian provinces are in the dark when power fails in an Ontario power plant.	• The U.S. Civil Rights crisis heats up, with the murder of Malcolm X and the march on Selma. • United States sends troops to the Dominican Republic. • Medicare is established as a medical insurance program for U.S. seniors. • The first riots in Watts, Los Angeles, California occur.
1966		• The second riots in Watts, Los Angeles, California occur.
1967	• Three U.S. astronauts are killed in a fire on the launching pad. • China tests its first hydrogen bomb. • The first successful heart transplant occurs.	• The Arab-Israeli Six-Day War ends with an Israeli victory. • Racial unrest spreads to many major U.S. cities.
1968	• Douglas Engelbart invents the computer mouse. • Bob Noyce and Gordon Moore found Intel. • Canadians Graeme Ferguson, Roman Kroitor, and Robert Kerr invent the IMAX movie system.	• North Korea seizes a U.S. naval vessel, accusing its crew of spying. • Vietnam War heats up with the Tet offensive and the Mai Lai massacre. • Martin Luther King, Jr. is assassinated. • The Soviet Union invades Czechoslovakia.
1969	• ARPANET, the U.S. government–sponsored network that eventually becomes the Internet, begins. • The first ATMs appear. • U.S. astronauts become the first men to walk on the moon.	• The Woodstock music festival takes place. • Richard M. Nixon begins his first term as U.S. president. • The gay rights movement begins with the Stonewall riot in New York City.
1970	• Alan Shugart invents the floppy disk. • Daisy-wheel printers are invented. • The first pocket calculators are sold. Unlike earlier electronic calculators, these run on batteries.	• Biafra loses its war of independence from Nigeria. • Rhodesia leaves the British Commonwealth. • The United States invades Cambodia. • Students protest the Cambodian invasion. At Kent State University in Ohio, four students are killed by the National Guard during a demonstration.
1971	• Pocket calculators are sold throughout the world, although they are still very expensive. • Dot-matrix (impact) printers are invented. • James Fergason invents LCDs.	• Busing of school children is approved in the United States as an aid to ending segregation in public schools. • The United States lowers its voting age to 18.

(continues)

Table 2.5 Technology and Historical Developments During the Second Half
of the 20th Century (continued)

Year	Major Events in the Development of Technology	Major Events in World History
1971 (cont.)	• Frederico Faggin, Ted Hoff, and Stan Mazor, working for Intel, develop the first microprocessor, the Intel 4004. • The first videocassette recorder appears. • Steve Jobs and Steve Wozniak begin selling the Apple I computer as a kit to hobbyists.	• Communist China becomes a member of the United Nations; Nationalist China (Taiwan) is expelled.
1972	• Intel releases the 8008 microprocessor. • The first word processor appears. • Pong, the first video game, appears.[7]	• Great Britain takes over Northern Ireland. • The Watergate break-in occurs. • Arab terrorists kill eight Israeli athletes at the Munich Olympics.
1973	• Robert Metcalfe invents Ethernet, the type of local area networking most commonly used today.	• The U.S. Supreme Court rules that women have a right to abortions. • The Vietnam War ends. • Nixon accepts responsibility for the Watergate break-in. • Greece becomes a republic. • The Yom Kippur War between Israel and its Arab neighbors ends with an Israeli victory.
1974	• Intel develops the 8080 microprocessor. It is used in the Altair, the first microcomputer. • Bill Gates and Paul Allen found Microsoft.	• U.S. President Nixon resigns and is pardoned immediately by his successor, Gerald Ford.
1975	• The first laser printer appears. • A U.S. spacecraft links up in orbit with a Soviet spacecraft.	• Cambodia is taken over by Communist forces (Pol Pot and the Khmer Rouge).
1976	• The first ink-jet printer appears. • Apple Computer is formed by Steve Jobs, Steve Wozniak, and Mike Markkula.	• U.S. Supreme court rules that minorities may be given retroactive job seniority. • Israel frees 103 hostages being held on an airplane in Entebbe, Uganda, by pro-Palestinian extremists.
1977	• The Apple II computer appears. • The Tandy and Commodore companies are founded and sell microcomputers. • Artificial insulin is first made.	• U.S. President Carter pardons Vietnam-era draft evaders. • A nuclear proliferation treaty is signed by the United States, the Soviet Union, and 13 other countries.
1978	• VisiCalc, the first electronic spreadsheet, is written by Dan Bricklin and Bob Frankston. This program will validate the microcomputer as a business machine. It will spur sales of the Apple II and make it clear to IBM that they are missing out on an important market. • Intel releases the 8088 microprocessor, which is later used in the original IBM PC.	• Rhodesia begins the change to black-majority rule. • The United States agrees to turn the Panama Canal over to Panama by the year 2000. • The U.S. Supreme Court bans quotas in college/university admissions but allows preference to be given to minorities. • The Camp David Accords end with a peace plan from Egypt and Israel.

[7]You can find a shareware version of the original Pong at http://www.freedownloadscenter.com/Games/Classic_Games/Pong.html

Year	Major Events in the Development of Technology	Major Events in World History
1979	• The Stanford Cart navigates its way across a room, demonstrating the first use of robotic vision. • Seymour Cray develops the Cray supercomputer. • The Walkman first appears. • The Three Mile Island nuclear plant suffers a partial meltdown, although the release of radiation is minimal.	• The Shah of Iran is ousted, leaving room for the ascendancy of fundamentalist Islamic rule. • Iranian militants swarm the U.S. Embassy in Tehran and take hostages. • Margaret Thatcher becomes Prime Minister of Great Britain. • The Soviet Union invades Afghanistan.
1980		• The hostages remain in Tehran. • The Iran-Iraq War begins; it will end in 1986. • Ronald Reagan becomes U.S. President. • John Lennon of the Beatles is killed in New York City.
1981	• MS-DOS is written. • The IBM PC (**Figure 2.14**) is released. Business users rejoice because "no one ever got fired for buying IBM." The machine sells like crazy. • Adam Osborne releases the Osborne I, the first truly portable computer. • Commodore introduces its most successful machine, the VIC-20.	• U.S. hostages are freed in Tehran. • AIDS is identified as a disease.
1982	• The first personal computer virus, Elk Cloner, appears. Begun as a ninth-grade student prank, the boot sector virus displays a poem on the computer's screen every 50th time an infected floppy disk is used to boot the machine. • Intel releases the 80286 microprocessor. • Commodore releases the C64.	• The British defeat Argentina in the Falkland Islands. • Israel invades Lebanon in an attempt to contain the Palestine Liberation Organization (PLO).

(continues)

Figure 2.13 An IBM 360, circa 1964.

Figure 2.14 The original IBM personal computer.

Table 2.5 Technology and Historical Developments During the Second Half
of the 20th Century (continued)

Year	Major Events in the Development of Technology	Major Events in World History
1983	• Apple releases the Lisa, the precursor to the Macintosh. Its operating system features one of the first graphic user interfaces. Unfortunately, the machine is too expensive and too slow to do well in the marketplace. • AT&T begins selling the first cell phones. • The first permanent artificial heart is implanted.	• A South Korean civilian airliner accidentally violates Soviet airspace and is shot down. There are no survivors. • Terrorists target a U.S. Marine barracks in Beirut, killing 237.
1984	• The Apple Macintosh debuts. Although it was first thought to be too cute to be taken seriously as a business machine, the graphic user interface made practical by its operating system becomes the basis of a major paradigm shift in computing. • Michael Dell founds Dell Computing. • IBM releases the PC AT. • The space shuttle *Challenger* completes its maiden voyage successfully.	• The Bell telephone system is broken up into AT&T and a number of regional service providers. • United States removes all troops from Beirut and warships from the surrounding waters. • The Soviet Union boycotts the Olympic Games held in the United States. • Indira Gandhi, the Prime Minister of India, is assassinated. • A Union Carbide plant in Bhopal, India releases toxic gas. Over 1,000 are killed and 150,000 injured.
1985	• Microsoft begins work on Windows. • Intel releases the 80386 microprocessor. • Ted Watt founds Gateway. Eventually, the company sells computers under the eMachines brand. • Aldus releases PageMaker for the Macintosh, the first desktop publishing software. This software begins a major paradigm shift in the publishing industry.	• PLO extremists hijack the cruise ship *Achille Lauro*, killing one person. • Airline hijacking becomes a significant world problem, with hijackings in Italy, the United States, and Egypt.
1986	• The *Voyager 2* spacecraft reports back from Uranus.	• Philippine President Marcos is ousted and replaced by an elected president, Corazon Aquino. • Halley's Comet returns. • The Chernobyl nuclear plant in the Soviet Union suffers a major meltdown, releasing significant amounts of radiation. • The United States enacts stiff sanctions against segregated South Africa. • The United States reveals that it has secretly been selling arms to Iran and sending the profits to the Contra rebels in Nicaragua.
1987	• IBM releases the PS/2 family of computers. • PageMaker becomes available for the IBM PC.	• U.S. hearings reveal the depth of the Iran-Contra scandal.

Year	Major Events in the Development of Technology	Major Events in World History
1988	• Digital cell phones are introduced. • Christian Andreas Doppler invents Doppler radar. • A patent is issued for the first genetically engineered animal.	• United States and Canada sign a free trade agreement. • U.S. Navy ship mistakes Iranian airliner for a jet fighter and shoots it down, killing all 290 aboard. • Benazir Bhutto becomes the first Muslim prime minister of Pakistan. • A terrorist bomb destroys a Pan Am 747 over Locherbie, Scotland, killing 270 people.
1989	• The concept of high-definition television is revealed. • Intel releases the 80486 microprocessor. • The oil tanker Exxon *Valdez* ruptures in Alaska's Prince William Sound, creating the worst oil spill in history. • The Voyager 2 spacecraft flies by Neptune.	• Chinese students and dissidents take over Bejing's Tianamen Square. The government uses tanks to disburse the gathering, resulting in a significant loss of life. • The Berlin Wall comes down and people move freely between East and West Berlin. • Czechoslovakia throws the Communists out of its government. • Romania throws the Communists out of its government. • The United States invades Panama to capture Manuel Noriega, the leader whom the U.S. government likens to a gangster.
1990	• Tim Berners-Lee develops HTML and creates the World Wide Web. • Microsoft releases Windows 3.0, the first truly usable version of Windows. However, at this point it is not a standalone operating system but a user interface shell that runs on top of MS-DOS. • The world's first search engine, Archie, becomes available. • Gopher, a graphic search tool, debuts. • The Hubble Space Telescope is launched. • IBM introduces its System/390 line of mainframes, including the System/9000 family.	• Yugoslavian Communists relinquish sole government power. • Soviet Communists relinquish sole government power. • South Africa frees Nelson Mandela. • Iraq invades Kuwait. A coalition of Western and Arab forces begins the Persian Gulf War to restore Kuwaiti independence. • East and West Germany are reunited into a single country, Germany.
1991	• Linux Torvalds releases the first version of Linux. • The first digital answering machine appears. • PGP (Pretty Good Privacy) debuts.	• Persian Gulf War ends. • France signs antinuclear proliferation treaty. • Albania's Communist government is replaced peacefully. • Lithuania, Estonia, and Latvia become independent nations. • Israel and Russia reestablish a relationship with each other. • The Soviet Union falls apart.

(continues)

Table 2.5 Technology and Historical Developments During the Second Half
of the 20th Century (continued)

Year	Major Events in the Development of Technology	Major Events in World History
1992	• Grace Hopper, generally regarded as the world's first programmer, dies.	• The Yugoslav Federation breaks up into separate nations, including Bosnia, Serbia, and Croatia. "Ethnic cleansing," the genocide of Muslims by Christians, begins. • The United States lifts trade sanctions against China. • The North American Free Trade Agreement (NAFTA) is presented. • The United States pulls all remaining troops out of the Philippines. • Czechoslovakia breaks into two nations, the Czech Republic and the Slovak Republic.
1993	• Intel releases the first Pentium processor. • Microsoft releases Windows NT. • The Motorola PowerPC chip appears. It will be used in Macintosh computers until 2006.	• Besieged Bosnian towns are supplied by airlift. • U.S. Federal agency raids the compound of religious extremists in Waco, Texas. All inside are killed by fire. • Civil War in Somalia claims the lives of U.N. troops. • Iraq allows U.N. weapons inspectors into the country. • China conducts a nuclear weapons test. • The European Union is created by treaty. • NAFTA is approved by the United States. • South Africa adopts a constitution that supports majority rule. • A terrorist bomb explodes in a parking garage of the World Trade Center in New York City. The structure of the building is not seriously compromised.
1994	• Marc Andreesen and James H. Clark found Netscape. • IBM releases OS/2 Warp. Although a solid operating system, OS/2 never makes significant inroads against Windows. • Iomega releases the Zip drive. • Red Hat Linux is founded. • Intel is forced to recall some Pentium processors because there is an error in the floating point unit that may generate incorrect answers to mathematical operations. • Yahoo is founded.	• Fighting continues between the nations of the former Yugoslavia. • Civil war in Rwanda begins, involving the genocide of Tutsis by Hutus. • Nelson Mandela becomes president of South Africa. • Jean-Bertand Aristide returns to Haiti and forms a new government. • Chechnya attempts to secede from Russia; Russia sends in troops. • Although the U.S. Supreme Court reaffirms a woman's right to choose an abortion, antiabortion extremists kill workers at clinics that perform abortions. U.S. Congress vows to protect the clinics.

Year	Major Events in the Development of Technology	Major Events in World History
1994 (cont.)	• Microsoft releases Windows 3.11. (It's still not a complete operating system.) • Commodore declares bankruptcy. • The Java programming language is developed by Canadian James Gosling.	
1995	• DVDs first appear. • eBay debuts. • Apple releases FireWire. • The first USB standard appears. • Intel releases the Pentium Pro processor. • Microsoft releases Windows 95, the first release of the software to be a standalone, complete operating system. • Amazon.com is founded.	• Mexico's troubled economy gets a boost from $20 billion in aid from the U.S. government. • Terrorists attack a Tokyo subway with nerve gas, killing eight and injuring thousands. • A terrorist car bomb explodes at the Oklahoma City Federal Building. • The genocide in Rwanda continues. • Fighting in Bosnia and Croatia comes to an end with a cease-fire. • France tests a nuclear device. • The Israelis agree to give Palestinians control of the West Bank of Jordan. The status of Jerusalem, however, remains in limbo. • Quebec's attempt to become independent of Canada fails by a narrow vote margin. • Nigeria hands power to minority rights activists.
1996	• World chess master Garry Kasparov is beaten by IBM's Big Blue computer. • Windows CE is released. • WebTV debuts. • Sergey Brin and Larry Page begin work on Google.	• France indicates that it will do no more nuclear testing. • China joins a global ban on nuclear testing. • Mad cow disease is diagnosed in Great Britain. • Russia signs a peace treaty with Chechnya. • A bomb explodes at an entertainment event during the Atlanta, Georgia Olympics. • Iraq attacks its Kurdish population; the United States attacks Iraqi air defenses. • Kabul, Afghanistan is captured by the Taliban. • There is violence in Zairian refugee camps, but in Rwanda and Burundi, refugees leave camps and return home.
1997	• Intel releases the Pentium II processor. • The 802.11 standard for wireless networks debuts. • A U.S. space shuttle docks with a Russian space station. • An unmanned U.S. spacecraft sends back pictures from Mars.	• Israel returns most of Hebron on the West Bank to the Palestinians but allows an Israeli settlement in East Jerusalem. • Albanian economy is in turmoil because of pyramid schemes. • U.S. Senate approves a chemical weapons treaty. • U.S. unemployment rate is the lowest in 24 years. • The European Union approves the adoption of the Euro as a common currency. • Iraq kicks out the U.N. weapons inspection team.

(continues)

Table 2.5 Technology and Historical Developments During the Second Half
of the 20th Century (continued)

Year	Major Events in the Development of Technology	Major Events in World History
1998	• Apple introduces the first iMacs. • PayPal debuts. • Google debuts. • Windows 98 is released. • Intel releases the Celeron processor. • DEC's assets are sold to Compaq.	• United States gets its first balanced budget in 30 years. • Fighting erupts in Kosovo (Serbs against ethnic Albanians). • Ted Kaczynski (the "unabomber") is sentenced to life without parole. • Russian economy nears collapse. • United States has a budget surplus. • United States conducts air strikes against Iraq, which has stopped cooperating with U.N. arms inspections. • U.S. President Clinton impeached but not removed from office.
1999	• Intel releases the Pentium III processor. • The 802.11b standard for wireless networks is released by the IEEE (Institute of Electrical and Electronic Engineers). • The judge in the U.S. government's antitrust case against Microsoft finds that the company is a monopoly. • The first Chinese spacecraft is launched.	• NATO planes bomb Serbian targets in an attempt to stop the violence in Kosovo. Serbs eventually agree to pull out of Kosovo. • Chinese government bans the Falun Gong sect. • Dagestan, a region of Russia, declares itself as an independent Muslim nation. • East Timor becomes independent of Tunisia. • Russia sends troops into Chechnya in renewed fighting. • Pakistani government is overthrown in military coup. • Northern Ireland begins self-rule. • The United States enacts a trade agreement with China.

2000 AND BEYOND

The 21st century is still very young, but technologically it feels somewhat different from the preceding decades. We are nearing the limits of how far we can push the speed of silicon-based microprocessors, the idea that oil is a finite resource is beginning to be taken seriously, and the planet is warming. The comfort that our technology brought us in the past 30 or 40 years can no longer be taken for granted: Unless we make major changes to our technology, Western societies cannot sustain their current standard of living.

Table 2.6 shows the efforts to create alternative technologies beginning to appear. Although history seems to be repeating itself, technology may be heading in new directions.

Table 2.6 Technology and Historical Developments During the First Decade of the 21st Century

Year	Major Events in the Development of Technology	Major Events in World History
2000	• The Y2K bug is a non-issue. • Windows 2000 and Windows ME are released. • Intel releases the Pentium 4 processor. • The United States passes the Children's Online Privacy Protection Act in an attempt to protect children under 13 from Internet predators.	• Peace accord signed between North and South Korea. • U.S. stock market falls significantly, primarily because of a large number of Internet business failures. • Israel removes troops from southern Lebanon, where they had been for 22 years. Violence continues elsewhere in the region. • Yugoslav president Slobodan Milosevic is thrown out after refusing to concede an election he lost; he is later tried by the United Nations as a war criminal for his role in the persecution of Muslims in the region.
2001	• Macintosh OS X debuts, followed later in the year by a second release. • The Apple iPod debuts. The MP3 player's user interface gives it an enormous edge over other players on the market. Unfortunately, Windows users will have to wait, because it works only with the Macintosh. • The AbioCor artificial heart, the first self-contained artificial heart, is used successfully. • An implantable drug pump is approved for use by patients who need continuous, automatic infusions of drugs. • A shirt that monitors the wearer's heart and respiration rates, body temperature, and calorie use is developed. • An artificial liver, that works much like a kidney dialysis machine, is developed. • Ford debuts an electric car. It can go only 55 miles on a charge and never reaches a mass market. • Aprilla debuts an electric bicycle that is powered by a hydrogen fuel cell. • The Segway Personal Transporter (**Figure 2.15**) is demonstrated. It is a self-propelled device that is controlled by the body position of the rider. • Buses that drive themselves by watching a line painted in the street debut in Las Vegas. • Digital satellite radio debuts. • First Alert markets a smoke alarm that can be turned on and off with a remote control.	• U.S. spy plane collides with a Chinese spy plane over China; crew is returned when U.S. states its regret over the incident. • The United States refuses to sign a global warming accord, which is signed by 178 other nations. • Ethnic fighting in Macedonia stops after 6 months. • Terrorists crash two passenger jets into the twin towers of the World Trade Center in New York City and a third into the Pentagon. There are more than 3,000 dead. Osama bin Laden and his al-Qaeda terrorist group are found to be responsible (referred to as the 9/11 bombings). • Anthrax kills five people after those people handle letters contaminated with the disease. • The United States and Britain begin bombing of al-Qaeda and Taliban targets in Afghanistan in retaliation for the 9/11 bombings. The Taliban loses control of the Afghan government.

(continues)

Table 2.6 Technology and Historical Developments During the First Decade
of the 21st Century (continued)

Year	Major Events in the Development of Technology	Major Events in World History
2001 (cont.)	• PPG Industries brings SunClean glass to market. This glass virtually cleans itself. • The Steri-Pen, which removes 99 percent of bacteria and pollutants from 16 ounces of water, debuts. Freeplay Energy develops a hand-cranked power source for a cell phone. • A Nigerian pot maker develops a double-pot system for cooling food without electricity. • A garden robot that eats slugs is developed. • Windows XP is released.	
2002	• Apple releases software to support the iPod under Windows. • The birth control patch debuts. • Developers show a telephone that can be embedded in a tooth. • A scramjet (a nonpolluting jet engine) makes its first flight outside a wind tunnel. • Wireless Bluetooth headsets appear. • A coaster that can detect date rape drugs is distributed. You put a drop of a drink on the coaster and if it turns blue, the drink has been spiked. • A military robot is designed to find land mines and use water to disarm them without exploding them. • iRobot's Roomba, the robot vacuum, appears.	• Twelve European countries begin using the Euro. • United States and Afghanistan launch a ground attack against remaining Taliban and al-Qaeda. • United States and Russia agree to downsize their nuclear arsenals. • U.S. President Bush begins speaking about problems with Iraq. U.N. resolution demands that Iraq disarm. Later in the year, Iraq allows weapons inspectors to return. • North Korea violates treaty restrictions and develops nuclear weapons. • United States backs Israel in its call for the ouster of Palestinian leader Yasir Arafat, who Israel labels a terrorist.

Figure 2.15
The Segway Personal
Transporter.

Year	Major Events in the Development of Technology	Major Events in World History
2002 (cont.)	• Robotics further invades toys with the debut of Cindy Smart Doll, which is equipped with voice recognition and robotic vision. • Second Life, the three-dimensional virtual world on the Web, debuts. • The world's most powerful supercomputer is built in Japan. It can perform ~35 million calculations per second and is used to create a climate model of the earth to predict weather and the effects of global warming. • Canesta and VKB demonstrate virtual keyboards that use lasers to detect finger movements without physical keys. • Hewlett-Packard buys Compaq.	• Corporate scandals rock U.S. business, including Enron (mishandling of funds) and WorldCom (misstating profits), The Enron collapse costs thousands of people their pension funds.
2003	• iTunes becomes the most successful place to legally purchase and download music. • MySpace debuts. • Flu shots can be delivered as a nasal mist. • A specialized microchip that can store an entire human's genome appears. It is useful when doing testing before genetic counseling. • A Japanese researcher develops a reflective material that makes the wearer appear to disappear. • Explosion of the space shuttle Columbia occurs.	• U.S. budget runs a large deficit. • U.N. weapons inspectors find no "weapons of mass destruction" in Iraq. Nonetheless, U.S. President Bush announces his intent to attack Iraq. U.S. Secretary of State attempts to justify a U.S. attack to the United Nations; Security Council members are not impressed. • One month later, the United States invades Iraq. Although official fighting ends within three weeks, guerilla and terrorist activity continues. The Iraqi leader, Saddam Hussein, is captured by the United States in December. • The European Union adds 10 more members. • Iran found to be conducting nuclear power/weapons development. • U.S. Supreme Court allows affirmative action in college/university admissions. • The peace process collapses in the Middle East.
2004	• Adidas partners with Apple to produce a shoe that automatically adjusts the firmness to the needs of the wearer. • The U.S. Mars rovers land successfully and begin transmitting images back to earth. Both rovers will last much longer and travel much further than NASA had anticipated.	• Chief U.N. weapons inspector states that there are no weapons of mass destruction in Iraq. • Al-Qaeda sponsors terrorist attacks in Spain. • NATO admits new countries that became independent after the breakup of the Soviet Union. • Rebels in Sudan reach agreement with the Sudanese government, but genocide in the Darfur region continues. • The torture of Iraqi prisoners by U.S. solders at the Abu Ghraib prison is revealed.

(continues)

Table 2.6 Technology and Historical Developments During the First Decade of the 21st Century (continued)

Year	Major Events in the Development of Technology	Major Events in World History
2005	• YouTube debuts. • Facebook debuts. • IBM sells its PC division to Lenovo, a Chinese firm. • Mac OS X 10.5 ("Tiger") is released. • Apple announces that it will migrate the Macintosh from Motorola PowerPC CPUs to Intel CPUs. • A flaw in CardSystem's security makes 40 million credit card numbers readily available to hackers.	• U.S. war in Iraq continues with troops fighting terrorist and guerilla forces. • Syria removes its troops from Lebanon after 29 years. • Terrorist bombings in London kill more than 50. • The United States signs the Central American Free Trade Agreement. • Civil war ends in Indonesia. • Israel begins removing settlers from the Gaza Strip as a prelude to returning the land to the Palestinians. • Saddam Hussein is tried for murder in an Iraqi court.
2006	• The Blu-ray technology for high-definition DVDs is developed. • Intel ships its Core Duo and Core 2 Duo processors. • Microsoft releases its Zune media player in an attempt to compete with the iPod. It proves to be too little, too late. • Pluto is reclassified as a dwarf planet; our solar system now has eight major planets and three dwarf planets. Many other rocks in space may eventually be labeled as dwarf planets. • First Intel-based Macintoshes ship.	• U.S. war in Iraq continues, with troops fighting terrorist and guerilla forces. • Iran restarts its nuclear program, claiming the work is for peaceful uses only. • Saddam Hussein is charged with genocide. He is convicted and executed. • Montenegro votes for independence from Serbia. • Israel sends troops back into Lebanon to fight Hezbollah, a Lebanese militant group. • North Korea conducts an atmospheric test of a nuclear weapon.
2007	• The Apple iPhone is released. One million units are sold in 74 days. • Windows Vista appears. • Intel releases the Core 2 Quad processor.	• U.S. war in Iraq continues, with troops fighting terrorist and guerilla forces.

WHERE WE'VE BEEN

Humans began using tools to help with computations in prehistoric eras. As people learned to shape metal, they created mechanical devices for computation. Even the first electronic devices, which used vacuum tubes, were designed for manipulating numbers: cracking enemy codes, aiming guns, and tallying a census. As vacuum tubes were replaced by integrated circuits, the use of computers expanded to include text, graphics, audio, and video.

The basic trend in computing throughout history has been "faster, smaller, and cheaper." This trend continues today with the introduction of smaller devices with increased functionality.

THINGS TO THINK ABOUT

1. The Osborne Computer Company made computers that "satisficed," Adam Osborne's term for something that met a user's needs but that wasn't necessarily cutting edge. What do you think of this strategy? Is it implemented anywhere today? Would you purchase a computer that was attractively priced but didn't have all the latest features? Why or why not?

2. Over the years, a number of computer vendors—Compaq, Commodore, DEC, for example—have been established and then dropped out of the marketplace. Why do you think this happened? Is it unique to the computer industry?

3. Some people think that wars stimulate technological growth. Looking at the technological advancements and major world conflicts in this chapter, do you agree with that assessment? If you do, why do you think this occurs?

4. Consider the places where technological development has occurred. Before World War II, where did most developments occur? What has happened since World War II? Why do you think this is so?

5. A spreadsheet developed by the Lotus Corporation (1-2-3) came to dominate the spreadsheet market in the 1980s. At one point in its history, Lotus purchased the company that developed VisiCalc, the first spreadsheet. Soon afterward, Lotus discontinued VisiCalc, leaving the field free for 1-2-3. This type of activity has happened several times. For example, Adobe, which developed the drawing program Illustrator, purchased Macromedia and eventually discontinued Macromedia FreeHand, Illustrator's greatest competitor. What do you think of this type of business activity? Is it unethical, or is it simply good business Why?

WHERE TO GO FROM HERE

Abbate, Janet. *Inventing the Internet*. Cambridge, MA: The MIT Press, 2000.

Allan, Roy A. *A History of the Personal Computer: The People and the Technology*. Hersham, United Kingdom: Allan Publishing, 2001.

Berners-Lee, Tim. *Weaving the Web: The Original Design and Ultimate Destiny of the World Wide Web*. New York, NY: Collins, 2000.

Campbell-Kelly, Martin. *From Airline Reservations to Sonic the Hedgehog: A History of the Software Industry*. Cambridge, MA: The MIT Press, 2004.

Ceruzzi, Paul E. *A History of Modern Computing*. 2nd ed. Cambridge, MA: The MIT Press, 2003.

"A history of information technology and systems." *http://www.tcf.ua.edu/AZ/ITHistoryOutline.htm*

"History of science and technology." The Internet Public Library. *http://www.ipl.org/div/subject/browse/hum30.03.80/*[8]

"History of technology." Wikipedia. *http://en.wikipedia.org/wiki/History_of_technology*

Ifrah, Georges. *The Universal History of Computing: From the Abacus to the Quantum Computer*. Hoboken, NJ: Wiley, 2002.

"Internet resources for history of science and technology." University of Delaware Library. *http://www2.lib.udel.edu/subj/hsci/internet.htm*[9]

"Nineteenth century inventions 1800–1850." About.com: Inventors. *http://inventors.about.com/library/weekly/aa111100a.htm*[10]

Ornstein, Severo M. *Computing in the Middle Ages: A View from the Trenches 1955–1983*. Bloomington, IN: 1st Books Library, 2002.

Rojas, Raul, and Ulf Hashagen. *The First Computer—History and Architectures*. Cambridge, MA: The MIT Press, 2002.

Schoenherr, Steven. "Recording technology history." *http://history.sandiego.edu/GEN/recording/notes.html*

Williams, Michael R. *A History of Computing Technology*. 2nd ed. Los Alamitos, CA: Wiley-IEEE Computer Society Press, 1997.

---

[8]This page has a great selection of links to technology history sites.
[9]More links to history of science and technology pages.
[10]You can use the links on this page to reach times from prehistory through 2000.

Technological Failures 3

INTRODUCTION

For the most part, people in Western societies believe their lives have been enriched by technology. But we can only benefit from technologies that succeed in providing whatever service they were designed to provide and that are relatively safe. One of the biggest problems we have with new technologies is foreseeing their impact before they are put into widespread use; the result has been the failure (in sometimes a spectacular way) of what was once a promising technology.

Given that most of this book discusses technological success stories, in the name of fairness we should look at some technological failures: technologies that did not live up to their promise and that we have either abandoned or curtailed the use thereof.[1]

At the end of the chapter, we look at technological failure caused by breaches of security. We will also discuss the vulnerability in which our reliance on technology places us.

LIGHTER-THAN-AIR CRAFT

During the era between World War I and World War II, the state-of-the-art in air travel was the lighter-than-air craft (what we now call a "blimp"). A gondola hung beneath the gas bag to carry passengers and cargo; other passenger and crew accommodations might be within the gas bag itself. Aviation at that time was focused on using such airships for commercial travel.

The largest of the airships was the *Hindenburg* (LZ-129), a German passenger craft. Much larger than one of today's commercial airplanes, the *Hindenburg* provided sleeping cabins for passengers. Although only 78 × 66 inches, the cabins were certainly larger than what you get in a modern airplane.

The *Hindenburg* was an alternative to sailing on a steamship to cross the Atlantic. Passengers could relax on the promenade deck (**Figure 3.1**), use a quiet reading and writing room, socialize in the lounge, or use a smoking room. The dining room was large enough to seat 50 people at one time. A full complement included 50 passengers and 40 crew members.

For a time, the future of luxury travel aboard lighter-than-air craft seemed assured—until May 6, 1937. The *Hindenburg* was approaching a mooring at the Naval Air Station in Lakehurst, New Jersey. This was the airship's first trip to Lakehurst and many newsreel photographers and reporters were at the landing site to see it come in.

Although no one is quite sure what happened, the *Hindenburg* burst into flames about 295 feet from the ground. Unlike the blimps of today, the German airships of the 1930s were filled with hydrogen, a highly flammable gas. (The United States was refusing to sell Germany helium, the gas of choice.) Once the fire took hold, the *Hindenburg* exploded, killing 35 of the 97 people aboard as well as one member of the ground crew (**Figure 3.2**). (This particular trip carried 35 passengers and 21 extra crew members.) The disaster was well captured on film (both still and moving) because there were so many news people in attendance.[2]

The *Hindenburg* tragedy became known across the globe very quickly. People could see it happen in the newsreels at movie theaters. (Remember, there was no tele-

---

[1] It's not unusual for technologies to have problems. In July 2007, for example, the customs computer at Los Angeles International Airport went down for seven hours. Over 6,000 international travelers were stranded because they couldn't be processed through immigration. This isn't the type of failure we're talking about in this chapter. U.S. Customs isn't likely to abandon or curtail its use of computing because its system went down. In this case, the problem was fixed and the incoming passengers were handled. Here we're talking about far more catastrophic failures, those that made people think twice (and then again) about the safety of a technology.

[2] For more information about the Hindenburg and other lighter-than-air craft, see http://www.nlhs.com /index.html. You can watch the original newsreel film taken at the time of the crash at http://www.youtube.com/watch?v=8V5KXgFLia4

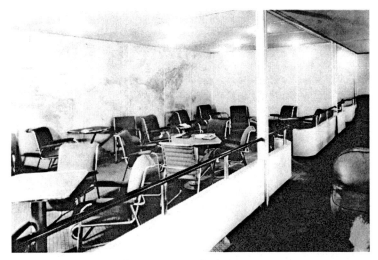

Figure 3.1 The *Hindenburg's* promenade deck.

Figure 3.2 The *Hindenburg* explodes on May 6, 1937.

vision in 1937!) The horror of the fire and explosion turned people throughout the world against lighter-than-air craft. In addition, commercial airplanes were becoming viable at that time. Shortly thereafter, humans abandoned lighter-than-air craft for personal travel. Today, the Goodyear Company, which began making blimps in 1925, operates only three lighter-than-air craft. All are filled with helium, which is nonflammable. These blimps are special-use vehicles, primarily for taking video and photographs from the air. None provides passenger service.[3]

---

[3] In-depth coverage of the Goodyear blimps can be found at http://www.goodyearblimp.com/basics/index.html

U.S. SPACE SHUTTLES

Since 1981 the space shuttle has been the cornerstone of the U.S. space program. Its major job has been to ferry large items into orbit, including the massive parts of the international space station. The shuttles, which will be retired in 2010, have transported countless satellites and the Hubble space telescope into orbit as well.

NASA's original intent for the space shuttle program was to have many flights each year. The shuttle was to be a reusable, freight truck in space. In fact, in 1985 there were nine shuttle missions; through the 1990s NASA launched an average of six shuttles each year. By all accounts, the space shuttle program was a rousing success.

However, there were a number of concerns about the technology that went into the shuttle, and two failures of that technology led to the loss of two shuttles and their crews. The first occurred on January 28, 1986. The shuttle *Challenger's* crew of seven included the first teacher to fly in space, Christa McAuliffe (**Figure 3.3**).

The presence of the teacher on board meant that arrangements had been made for school children across the country to watch the launch. The first part of the launch was successful, but 73 seconds into the flight, *Challenger* unexpectedly exploded in midair (**Figure 3.4**). Like the explosion of the *Hindenburg*, the explosion of the *Challenger* was caught on video by scores of media representatives.

The direct cause of the *Challenger* explosion was a technology failure. The segments of the shuttle's solid booster rockets, which were installed on either side of the external fuel tank, were held together by O-ring seals. In the below-freezing temperatures of a January morning, the seals failed, allowing flames from the rockets to cause a leak in the external fuel tank. The ignition of the liquid oxygen fuel in the external fuel tank then caused the explosion.

But the possibility that the O-ring seals would fail in cold temperatures was not unknown. In fact, a group of engineers who had been working on the shuttle warned shuttle program middle-level managers of the danger and urged them not to launch on that day. (In fact, the problem with the O-rings had been known since 1977, but nothing had been done to fix it.) There was, however, a great deal of pressure on top-level administrators to get the shuttle off the ground—the flight had been scrubbed four times before—and they decided to go ahead. Part of the problem was that the organizational structure at NASA made it difficult for the engineers to make themselves heard; the middle managers to whom they voiced their concerns did not send those concerns up the organization hierarchy.

The *Challenger's* crew cabin was not destroyed by the explosion. However, no provisions had been made for astronauts to exit the cabin in case of disaster. Although it can't be determined exactly when the astronauts died, it is certain that they had no chance to escape the cabin, which tumbled in free fall and ultimately hit the ocean at more than 300 kilometers per hour. The lack of escape provisions was a conscious decision on NASA's part, because they believed the shuttle to be so reliable that escape technology wasn't necessary.

The result of the *Challenger* disaster was a 32-month hiatus in shuttle flights while NASA attempted to address some of the shortcomings—both technologi-

Figure 3.3 The crew of the space shuttle *Challenger*. (Back row, from left to right: Ellison Onizuka, Christa McAuliffe, Gregory Jarvis, Judith Resnick; Front row, from left to right: Michael J. Smith, Dick Scobee, Ronald McNair.)

Figure 3.4 The space shuttle *Challenger* explodes.

cal and organizational—that were identified in the postexplosion investigations. However, they still chose not to include a crew cabin escape mechanism because they continued to believe the shuttle was quite reliable.

Shuttle flights resumed in 1988, when only two missions were flown. By 2000 NASA appeared to be quite ensured of the safety of its equipment.[4] Unfortunately, a second shuttle, *Columbia*, was lost on February 1, 2003. This time, the shuttle exploded on reentry rather than on launch.

The space shuttles are covered by thousands of heat-resistant tiles that allow the shuttle to withstand the heat of reentry. When *Columbia* launched, some insulation from the main fuel tank broke off and hit the tiles on the leading edge of the shuttle's left wing. Both the flight and ground crews knew the problem with the insulation had occurred but were unaware that tiles had come loose and that the tiles no longer provided the necessary protection from reentry heat. Temperatures during reentry rose, as was usual, to around 3,000°F. Because there was a break in the tiles' protective coating, the high temperatures caused the shuttle to break up before it was able to land.

After the loss of *Columbia*, NASA equipped the shuttles with a camera at the end of a long robot arm so that astronauts could inspect the tiles after launch. If a break in the tiles was found, astronauts could then plan a space walk to fix the problem before the shuttle attempted to land.

Like the problem with the O-rings, the problem with foam breaking off the fuel tanks was well known to NASA personnel. However, insulation had broken free so many times before with no ill effects that it had become commonplace and was therefore ignored. NASA regulations stipulated that flights were to be aborted when this occurred, but those regulations were routinely overlooked.

---

[4]On August 8, 2007, a teacher finally made it into space aboard the space shuttle Endeavor. Barbara Morgan, who was Christa McAuliffe's backup, made the flight. She had been training full-time as an astronaut since 1998 and had other responsibilities on board beyond those involving school children.

The space shuttle is a very complex piece of machinery. In some ways, it's surprising that it works at all! The technical problems that caused the loss of the two shuttles were known, but NASA's culture and organizational structure prevented action to solve the technical issues. The lesson here is that even if we can foresee the outcome or possible problems with a technology, we humans are still helpless to prevent the failure of that technology if cultural and social issues get in the way.

NUCLEAR POWER

Currently, we are in the midst of trying to effect a paradigm shift in the way in which we generate energy. Our reliance of dwindling supplies of fossil fuels has made us look at a variety of other technologies. One of those alternatives, nuclear power, initially showed a great deal of promise. This is another complex technology, the results of which became apparent only after the technology had been in use for a relatively long period of time. What makes the issues surrounding nuclear power so complex is that experts continue to disagree on the actual degree of risk, and therefore there are no solid answers to give to people who are concerned about the technology.

Current nuclear technology involves the release of energy when atoms are split into pieces (nuclear fission).[5] The result is used fuel rods and the radioactive results of the fission that must be stored for thousands of years. Operators of nuclear plants are therefore faced with the problem of finding stable containers and storage facilities. The fuel rods are typically stored in pools of boric acid, which helps the rods cool down. Storage in the pools is supposed to be short term (about 6 months), but because there is no long-term storage solution available, many rods stay in the pools for years.[6] This presents another problem: The rods must be kept away from one another so they don't start a nuclear chain reaction.

Fission products are typically buried in the ground. They must be stored in a dry location in containers that will not rust or rot over a period of thousands of years. One such burial site in the United States is in Yucca Mountain, a very dry region in the state of Nevada. A great deal of nuclear waste is also stored at the site of the Hanford, Washington nuclear plant. (Hanford no longer operates and is in the midst of a radiation cleanup; nonetheless, nuclear waste from all over the country is stored there.) However, very few locations want new nuclear storage in their backyard, so even though a site may be geologically acceptable, the residents of the area may prevent the establishment of the storage facility.

Waste storage problems are the result of nuclear power plants that work properly, but there is another danger that can occur when there is trouble in the reactor and/or its cooling systems: a core meltdown. A core meltdown begins when the cooling system is unable to successfully cool the core of the reactor; the

---

[5] An alternative technology, nuclear fusion, releases energy when atoms are combined. Nuclear fusion does not produce the radioactive waste associated with nuclear fission. As of this writing, no one has been able to make it work in a practical manner. For information on the state-of-the-art in cold fusion research, see http://www.wired.com/science/discoveries/news/2007/08/cold_fusion

contents of the reactor overheat and melt. The result can be a significant release of radioactive contamination into the air, with the associated risks for living creatures, including increased incidents of cancer (particularly of the lymphatic system) and birth defects.

Proponents of nuclear power stress the small chance of a core meltdown occurring. Nonetheless, there have been two well-known incidents, one in the United States (Three Mile Island) and one in the Ukraine (Chernobyl). The second was by far the more serious of the two.

Three Mile Island

The Three Mile Island nuclear site is located on an island near Middletown, Pennsylvania (**Figure 3.5**). It had two reactors, one of which is still operating.[7] The problem occurred on March 29, 1979 in the second reactor (TMI-2), the result of a series of technical failures.

First, the pumps that fed water to the generators that removed heat from the reactor stopped working. (No one is quite sure why the pumps failed, but it was some type of technical problem.) The reactor acted properly by shutting itself down, causing the pressure in the reactor to rise. A valve opened to release the pressure. At this point, the reactor was acting as it was programmed to act in the case of a cooling system failure. However, once the pressure in the reactor dropped to a given level, the valve should have closed: It didn't. Cooling water began to leak out of the valve and therefore the temperature of the core of

Figure 3.5
The Three Mile Island nuclear plant.

[6] It is possible to reprocess some of the contents of the used fuel rods so that they can be reused in power plants, thus reducing the amount of radioactive waste that needs to be stored. However, because the product of reprocessing is "weapons grade" nuclear material, the United States has banned the procedure for fear that the reprocessed material will fall into the wrong hands. This is just one example of a government curtailing the use of a technology for political reasons.

[7] The operating reactor's license expires in 2014, at which time the owner has indicated that it will be decommissioned.

the reactor began to rise. TMI-2 suffered a serious core meltdown. Fortunately, the containment building that held the reactor did not rupture or explode, and there was only a minimal release of radioactivity. Nonetheless, the fact that the meltdown occurred scared many people in the United States, especially because Three Mile Island is located near heavily populated areas.

Could the meltdown have been prevented? Perhaps, if the plant operators had recognized that the valve hadn't closed when it should and were therefore able to close it manually. But there was no single control room indicator that could tell the plant operators that the valve was still open. In addition, there was no indicator for the level of coolant in the reactor. (If there had been one, they could have seen the coolant level dropping and realized that the valve must have been open.) Therefore, even though alarms were going off throughout the plant, the operators weren't sure what was causing the emergency conditions. The actions they took actually reduced the level of coolant in the reactor and made the situation worse.

The meltdown at Three Mile Island was the result of two technical failures, but those failures could have been recognized and rectified before a meltdown if the indicators in the plant's control room had been clearer. This is a case of a complex technology where poor user interface design contributed to a potentially devastating accident.

Chernobyl

As far as the safety of the general public is concerned, the meltdown of the Three Mile Island reactor was more frightening than damaging, but the same is not true of the meltdown that occurred in the Chernobyl nuclear plant (near Pripyat, Ukraine) on April 26, 1986. The plant had four operating nuclear reactors, with two more under construction; the meltdown occurred in reactor 4, ultimately the result of miscommunication between plant operators as well as technical failure. **Figure 3.6** shows that the damage caused by the meltdown to just the reactor itself was considerable. And this doesn't reflect the problems caused by radiation to the surrounding area!

During the day of April 25, reactor 4 had been scheduled to run a test of the backup generators that would power its cooling system in the case of the loss of electrical power. To prepare for the test, the reactor's power output was reduced by 50% and needed to be reduced even further to perform the test safely. However, the person in charge of the electric supply for the city of Kiev said that reducing output further would not provide enough power for the evening peak demand. Therefore the test was postponed and delegated to the overnight shift.

The day shift neglected to tell the overnight shift how much the reactor output had already been reduced. The overnight operators therefore reduced the power output too quickly, resulting in production of an excess of xenon-135, a nuclear poison. The power output dropped more quickly than the operators expected, but the operators didn't recognize the cause. Instead, they thought that the problem was with one of the automatic power regulators, and they initiated the removal of control rods from the core, pulling more rods than was safe to remove.

Figure 3.6
The Chernobyl nuclear
plant after the meltdown.

At this point, the reactor's power output increased minimally but didn't reach the level needed for the test. Instead of aborting the test, however, the operators chose to continue. They increased the water flow to the reactor and pulled even more control rods. The reactor was then operating far outside its specified limits, but there were no indicators in the control room to alert the operators that the reactor had become unstable.

When the normal generators were disconnected to allow the test of the backup generators, the temperature in the reactor began to rise quickly. Steam bubbles formed in empty spaces in the coolant pipes. The excess xeonon-135 was used up and added more fissionable material to the core. Unfortunately, too many control rods had been removed from the reactor to contain the runaway nuclear fission.

The operators attempted to shut down the reactor and reinsert the control rods. The design of the reactor, however, was such that the result was the opposite of what the operators intended: The rods couldn't be inserted more than one-third of the way and the reaction in the core continued to speed up. There was nothing more the operators could do. The reactor suffered a major steam explosion that blew off the roof of the reactor's containment building, letting in oxygen that started a fire.

The explosion and fire spread radioactive contamination over a wide area near the plant. Although people were evacuated, more than 135,000 people living near the plant were subjected to high doses of radiation. Within 10 days, the radiation had traveled around the planet, although the worst of the devastation was within 100 miles of the plant. (The area of major contamination is not circular because of the wind patterns in the area.)

Of the 237 people who came down with severe radiation sickness immediately after the accident, 28 died shortly thereafter. These were mostly emergency personnel who were fighting the fire and trying to contain any residual radiation leakage. There is great controversy, however, over how the radiation has affected the people who were living in the area. An increase in treatable thyroid cancer in children has been documented, but it is still unclear whether there is an increase in birth defects or other cancers. (Leukemia cases may increase as the population grows older.) Animals have returned to the areas surrounding the nuclear plant and seem to be doing well in the absence of humans in the contaminated areas. On the other hand, it is difficult to determine the effects of the radiation because there is little information about "normal" cancer and birth defect rates in the area.

The area surrounding the Chernobyl plant is still uninhabitable by humans due to radiation, and it will not be usable again for thousands of years. Not only is the land contaminated, but so are groundwater supplies.

Who or what was responsible for the disaster at Chernobyl? One theory places all the blame on the plant operators who continued a risky test after they received abnormal readings from their indicators. Another theory believes there was a design flaw in the reactor containment rods and that the rods should have been able to contain the reaction, even when not fully inserted. It may well be that the accident was caused by a combination of both factors.

Future of Nuclear Technology

Despite its initial promise, nuclear power is out of favor with the people of North America. Few nuclear plants remain in operation; no new ones are being built nor have any been proposed. The problems associated with radioactive waste and the fears of another accident like that which occurred at Chernobyl have outweighed the benefits of generating large amounts of power without using fossil fuels.

Conversely, it is possible to look at the incident at Three Mile Island in terms of what *didn't* happen. A core meltdown occurred in a nuclear reactor, and only a minimal amount of radiation was released. One could say that the design of the reactor was so good that it prevented a tragedy, demonstrating that even under the worst of conditions, nuclear reactors are safe.

In contrast to the American fear of nuclear power, Europeans have welcomed it. For example, France gets three-quarters of its electricity from nearly 60 nuclear power plants. Germany gets about one-third of its power from nuclear energy (17 plants), and Sweden uses 10 nuclear generators to supply about half its power needs. Nuclear power has also worked extremely well on the water. The small power plants in nuclear-powered submarines and aircraft carriers have excellent safety records. One reason is that the military can enforce its procedures for operating the reactors more strictly than can a civilian organization. However, most Americans believe that nuclear power is a failed technology that should be abandoned, but despite the accidents that have occurred, not everyone in the world would agree.

DDT

DDT, the acronym for dichlorodiphenyltrichloroethane, is a pesticide that was originally synthesized in 1874. However, it wasn't until 1939 that scientists realized it could be used to kill certain insects. In particular, DDT is very good for killing the mosquitoes that carry malaria and typhoid, and in malaria-prone areas where DDT is used the incidence of malaria drops to nearly zero. DDT is also useful for killing a number of pests that attack farm crops, including flies, aphids, walking sticks, and Colorado potato beetles.

However, in 1962 the publication of a book, *Silent Spring* by Rachel Carson, alleged some very serious problems with DDT. In particular, its use was devastating some species of birds, such as the bald eagle, the brown pelican, and the California condor (**Figure 3.7**). The birds were laying eggs with thin shells that weren't hatching. The California condor population was already threatened, and the effects of DDT could have wiped it out entirely.

How in the world could a pesticide affect birds' eggs? The birds that were affected were those at the top of the food chain. Plants that had been sprayed with DDT were first eaten by small mammals. The concentration of DDT in the bodies of the mammals was therefore much higher than what was originally sprayed over large areas. The birds then ate the small mammals, further concentrating the DDT in their bloodstreams. The California condors were particularly vulnerable because they are scavengers and were as likely to eat the carcasses of birds of prey (such as the bald eagle) as anything else, further concentrating the pesticide.

The public outcry over the fate of the birds caused the United States to ban the use of DDT in 1972. Since then, most bird populations thought to be threatened by DDT have recovered; their eggshells have returned to an appropriate thickness within 20 years. The California condor, already endangered when DDT entered the picture (there were only 27 living condors when humans decided to help and at one point the population dropped to 5 wild birds) has not been as fortunate. There are fewer than 300 California condors alive today, most of which were bred in captivity and released into the wild in their original habitat areas (central

Figure 3.7
California Condor.

and southern California, Arizona, and Baja, Mexico).[8] The hopes of the captive breeding program—a joint effort between the United States and Mexico—is that the released birds will breed in the wild and slowly replenish the population.

Although most statisticians will tell you that "correlation does not imply causation," there is a very strong relationship between the use of DDT and the thickness of bird of prey eggshells. When DDT was removed from their environment, the birds recovered. For many people, that is enough to conclude that DDT was the cause of the birds' problems.

In the concentrations that were used on plants, DDT does not appear to be harmful to humans, and it continues to be used today in countries where malaria and typhoid are significant concerns. In the United States, however, it remains a banned substance, another technological product whose long-term effects the developers could not possibly foresee.

SECURITY BREACHES AND OUR RELIANCE ON TECHNOLOGY

If you ask an information technology professional about the most important issue in technology today, he or she will probably talk to you about security: Technology fails when security failures disclose private information or make technology unavailable.[9] A significant portion of corporate IT budgets goes toward security technology assets against both intentional and accidental security breaches.

Some security breaches have little more than nuisance value. For example, rival political groups have been known to deface Web sites of their opponents. Although ethically reprehensible, such activities rarely cause lasting harm or release private information to the public.

However, some security attacks pose significant risks. Consider, for example, what happened to Estonia in the spring of 2007, which is one of the most Internet-connected countries in the world. Its government and banking system rely heavily on the Internet to conduct business. Most Estonian citizens use the Internet to do their banking and to get current news. Estonia has also implemented online voting.[10]

In early April 2007 the Estonian government decided to relocate a statue commemorating Russian soldiers killed in World War II from a center-city venue to a rural cemetery. Some of the ethnic Russians who live in the country were disturbed by the relocation; some of Russia's leaders also expressed unhappiness with Estonia's actions.

Within a few days, Estonia's government Web sites (the presidency, the parliament, and government ministries) were the victims of what is known as a

---

[8]Reintroduction of the birds into the wild hasn't always gone smoothly. In the early years, five birds were killed when they collided with power lines. The young birds are now trained to avoid perching on power poles. For a description of the California Condor Recovery Program, see http://www.fws.gov/hoppermountain/cacondor/AC8&AC9.html

[9]Security and privacy are really two different things. "Privacy" refers to the need to keep certain data confidential. "Security" is the process of keeping private data private.

[10]For more information on Estonia's online voting system, see http://www.wired.com/politics/security/news/2007/03/72846

denial-of-service attack: The Web servers were flooded with so much traffic that they crashed, making the servers unavailable to legitimate users. The attacks spread to banking, newspaper systems, and communications firms.

The only way to stop the attacks, which came from outside the country, was to cut Estonia's connection to the Internet, isolating the country from the rest of the world. Although such an action did stop the attacks, it also prevented Estonia from letting others know what was happening. The country was effectively brought to a standstill. Fortunately, the attacks—which were largely automated—stopped after two weeks, and Estonia was able to restore its international Internet access.[11]

Estonia considered the attacks to be the equivalent of an invasion by a physical army. As a member of NATO, it could then call upon other NATO nations to help in its defense. However, at the time, NATO didn't recognize cyber-attacks as equivalent to military action and therefore as coming under its mutual-defense treaty provision.

Estonia implied that Russia was behind the attacks. Although officials did not blame Russia directly-doing so could have caused a dangerous breach between the two countries—the Estonians seemed fairly certain that Russians had hired malicious hackers who controlled networks of *zombie* computers, computers that had previously been infected by software that, when activated, would launch the denial-of-service attack.[12]

As frightening as the possibility of a cyber-attack may be, our technology is vulnerable to outside attacks at a much more fundamental level. Everything we do with technology relies on electricity, and most of the world gets its power from a local power grid. If the power grid goes down, much of our technology becomes unusable. Consider what happened to the northeastern United States and Canada on August 14, 2003 when there was a blackout that spread from New York north through Massachusetts, west to Michigan, and north to Ontario (**Figure 3.8**). Subways and other electric trains stopped running, elevators stopped working (sometimes between floors), traffic signals went dark, and lights in tunnels didn't function. Most telephones continued to operate, but cell phone towers were overloaded by increased traffic. Some municipal water systems were forced to shut down. This was more than an inconvenience; it became a public health and safety issue. By the time power was restored, the financial losses were estimated to be $6 billion.

The experience of the 2003 blackout taught us several important lessons. First, our power grid is probably our most vulnerable point. Without power, our technological civilization grinds to a halt. Second, we needed to build fail-safe controls into our electrical grid. The reason why such a large area was affected by the blackout was something known as the "cascade effect," where automatic

---

[11]There is no way to prevent the type of attack that was launched against Estonia. The only way to stop one is to do what Estonia did: break the connection.

[12]The zombies were located all over the world, many of them in the United States. How did the zombies become infected? The users of the machines opened e-mail containing the malicious software or visited Web sites that downloaded it to their computers. The users were probably totally unaware that their systems had been corrupted.

Figure 3.8 The area affected by the 2003 blackout.

attempts to provide power to an area without power caused further outages; one outage triggered another as power plants became overloaded. Finally, if terrorists wanted to cripple the entire North American economy, they could do so with concerted attacks on our power plants. We need to provide better security than was in place in 2003.

A person can react to technological failures of the type discussed in this section in a couple of ways. You can accept such failures as part of a technological society that is continuing to evolve and improve. Alternatively, you can decide that such failures are the reasons we should abandon technology and return to a simpler life.

MEDIA AND OUR PERCEPTIONS OF TECHNOLOGY

There is one persistent thread in all the technological failures you've read about so far: The United States has banned or abandoned a technology while the rest of the world continues to use it. Why might this be the case? Are Americans wiser than the rest of the world? Or are Americans more cautious, perhaps to a fault?

Part of the answer lies in the U.S. news media, and the propensity of reporters—print, audio, and video—to latch onto the negative aspects of technology and, in some cases, advocate for the abandonment of the technology. The news media have had a great influence on the American perception of various technologies, depending on the slant the reporter puts on the story.

For example, consider the meltdown at Three Mile Island. There are two ways to view that incident: either as a disaster that very nearly happened or the success of technology in preventing a disaster. The reporting of the incident was significantly slanted toward the former, emphasizing how much radiation *could* have been released and what might have happened had the containment building failed.[13] Depending on your point of view, such reporting is either a service to the people because it alerts them to a danger that needs to be handled or a negatively slanted take on the situation that suggests it was far worse than it was.

Most Americans don't want to muzzle the media, but we need to keep in mind that news reporting is not necessarily unbiased. Whenever we hear or read or view a news report about technology, we need to consider the type of report: If it is the result of "investigative reporting," for example, it will almost certainly have a bias. One of the most difficult challenges facing us today is sorting through the huge volume of media that we read, hear, and see to determine what is fact and what is someone's point of view. But if we are to evaluate reports of technological failures and their effects, we have to do just that: make judgments about what is true, what is good science, and so on.

PREDICTING THE CONSEQUENCES OF TECHNOLOGY

One of the reasons that technical failures can cause so much damage is that it is almost impossible to foresee all the consequences of using a technology. For example, when asbestos was developed, it appeared to be an effective fireproof insulation. No one was able to predict that inhaling asbestos fibers would cause cancer. Certainly, no one was able to predict the effect of DDT on birds of prey or the effects of lead paint on children.

Consider all the automobile recalls that we read and hear about. This does not suggest that the car is a failed technology but rather that no matter how carefully a car is tested, it is possible for mechanical and/or technological defects to remain. Sometimes the result is tragedy, such as when Ford Pinto gas tanks exploded after a rear-end collision.[14] In other cases, such as seatbelt buckles that come undone by themselves, the recall is ordered before there is widespread injury or loss of life.[15]

In the electronic realm, we are often faced with bugs in our software. The consequences of a software error can vary from the merely annoying—the loss of a document in progress—or catastrophic, where the inaccurate reporting of a sensor reading in a process control system that runs an oil refinery or nuclear plant causes loss of human life. Software developers put their products through rigorous testing programs and then usually send the software to beta testers, end users who put the software through heavy use, knowing significant bugs may still exist. As hard as they try, however, they rarely can find every problem. Users work

---

[13] For examples, see http://www.super70s.com/Super70s/News/1979/March/28-Three_Mile_Island.asp, http://www.eyeonwackenhut.com/index.asp?Type=B_BASIC&SEC=%7BC2969EF5-B919-451E-87C6 -082D39236A2F%7D, and http://web.mit.edu/newsoffice/2007/threemile.html

[14] As with many of the issues raised in this chapter, there are two sides to the story. Some people allege that Ford Motors knew about design flaws in the Pinto and chose to ignore them because it was cheaper to settle lawsuits than to redesign the car and order a recall of existing vehicles. Others note that the incidence of exploding gas tanks was very small (27 deaths from the over 2 million Pintos built) and that the car's safety record was comparable with other cars of its era.

[15] For an example, see http://www.lawyersandsettlements.com/case/defective-seatbelts-in-heavy-trucks.html

with software in ways the developers simply can't predict. Perhaps users press key combinations that the developer didn't anticipate, or perhaps users work with program features in an order that wasn't foreseen.

There is a story floating around cyberspace that when developing models of the Apple II computer, Apple testers would roll a glass Coke bottle around the keyboard to generate random key presses. They were well aware that users would do things on the keyboard that they couldn't possibly anticipate.[16] But not all hardware and software is amenable to the Coke bottle test. Predicting the consequences, outcome, or behavior of a new technology is one of the greatest difficulties we face when deciding which technologies we should use.

---

[16] Many stories abound, some of which are probably apocryphal, about the strange things that users do. For example, a user called a computer help line wanting to know why his computer's foot pedal didn't work. To that user, the mouse looked like a sewing machine foot pedal! Users reportedly also confused a CD-ROM tray with a cup holder. The classic story involves a user calling a help line, asking where he could find the "any" key on the keyboard. His software said "Press Any Key to Continue."

WHERE WE'VE BEEN

It can be very difficult to predict the long-term effects of a technology. Occasionally, those long-term effects can be negative, resulting in the loss of human life or serious injury. When such events occur, we tend to rethink the use of the technology that caused the accident. Sometimes the technology is abandoned or its use severely curtailed, especially when the failure is the result of a design problem with the technology itself. However, if the failure can be attributed to human error, we are more likely to adjust our behavior or to modify the technology to prevent the accident from recurring.

In some cases, a technological failure can be viewed in more than one way: Some people blame the technology itself, whereas others ascribe the failure to human error. The result is that uses of technologies that have exhibited major problems are not the same throughout the world. Some countries abandon the technology completely, whereas others continue to use it. One of the great influences in deciding which way people will view a technological failure is the media. The "spin" that media puts on an accident can significantly sway public opinion of the technology.

THINGS TO THINK ABOUT

1. Consider the drug thalidomide. What was its intended use? What were the problems with the drug? How serious were the consequences? In your opinion, do the benefits of the continued use of this drug outweigh the dangers? If so, what controls should be placed on its use to prevent problems from recurring?

2. Identify a technology that has had a failure reported by the news media. What slant has the media reporting placed on this failure? Has the reporting changed your view of the technology? If so, in what way?

3. Several of the technological failures discussed in this chapter were made worse by the failures of the organizations responsible for the technologies. Discuss some strategies that an organization can take to create an environment in which people are less likely to contribute to the effects of technological failure.

4. Some bugs are almost always going to remain in any piece of large software. Imagine that you are in charge of quality control/product testing for a software company. Prepare a testing plan that can minimize the number of bugs in your current project before it is released for general sale. (Hint: You can research software testing on the Internet.)

WHERE TO GO FROM HERE

Space Shuttles

"51-L: the *Challenger* accident." *http://www.fas.org/spp/51L.html*

"Space accidents." *http://www.infoplease.com/ipa/A0001458.html*

"Space shuttle *Columbia* and her crew." *http://www.nasa.gov/columbia/home/index.html*

Three Mile Island

"Fact sheet on the Three Mile Island accident."
   http://www.nrc.gov/reading-rm/doc-collections/fact-sheets/3mile-isle.html

"Three Mile Island emergency." *http://www.threemileisland.org/*

Chernobyl

"Chernobyl accident." *http://www.uic.com.au/nip22.htm*

"Chernobyl: a nuclear disaster." *http://library.thinkquest.org/3426/*

DDT

"DDT@3Dchem.com: banned insecticide." *http://www.3dchem.com/molecules.asp?ID=90*

"DDT banned insecticide." *http://www.3dchem.com/molecules.asp?ID=90*

"DDT ban takes effect" *http://www.epa.gov/history/topics/ddt/01.htm*

Edwards, J. Gordon, and Steven Milloy. "100 things you should know about DDT."
   http://www.junkscience.com/ddtfaq.html

Estonia

Smith, Adam. "Estonia: under siege on the Web."
   http://www.time.com/time/magazine/article/0,9171,1626744,00.html

Vamosi, Robert. "Cyberattack in Estonia-what it really means." *http://news.com.com/
   Cyberattack+in+Estonia-what+it+really+means/2008-7349_3-6186751.html*

Resisting Technology 4

WHERE WE'RE GOING

In this chapter we will

- Consider why humans are often resistant to change.
- Discuss the Luddites, who resisted technology during the Industrial Revolution.
- Look at the Amish, who resist some technologies while accepting others.
- Examine Greenpeace, an organization that resists technology that harms the environment.
- Discuss the Neo-Luddites and anarcho-primitivists, who advocate the dismantling of technological civilization.
- Talk about Ted Kaczynski, who took violent action against technology.
- Consider the irony of using the Internet to spread the word about resisting technology.

INTRODUCTION

Humans seem to have a natural resistance to change. We can come up with a wide variety of reasons why we don't want to disrupt the status quo:

- I'm happy with the way things are now. ("If it ain't broke, don't fix it.")
- I've put a lot of effort into getting where I am now and don't see any reason to go through the same effort again.
- I'm too busy.

- The result of the change doesn't seem to be an improvement. In fact, it seems likely to make things worse.
- The change doesn't particularly interest me.
- I don't know how to make the change happen, even though the plans look good.
- It is too much trouble to go through the change process. The result isn't worth the effort.
- I am able to stop the change, and I will because I can.

In addition to what we verbalize about change, we may also be afraid of it because no matter how good the plans, the result is unknown. Technological change is no different from any other type of change: Some people will resist it. Whenever we make a major technological change—such as replacing an existing coaxial cable network with a fiber-optic cable network—there's a leap of faith we have to make as we begin the project: We have to assume that the project will be successful and that the result will be an improvement over what we had previously. There is some level of risk associated with any change.

This chapter is about those who have resisted technological change. We start with the historical Luddites, those who—often violently—resisted the introduction of automated manufacturing. Then we will look at modern groups that oppose technology for a variety of reasons. One of the most interesting things about these groups and individuals—in particular, the historical Luddites and the Amish—is that although we often think of them as opposed to all technology, that is not necessarily the case.

HISTORICAL LUDDITES

Just before the Industrial Revolution, clothing was made by skilled craftsmen and craftswomen in their homes. These "cottage industries" produced a small quantity of high-quality goods. Because all the work was done by hand, prices were relatively high.

During the very early 1800s, English factories were established that used "stocking frames" to produce stockings. This machinery turned out stockings faster and for less money than handmade goods. The quality of the items was also inferior to those made by hand. Nonetheless, the lower prices for the machine-made stockings began to have a significant impact on the market for handmade goods, and the hand workers began to see a measurable decrease in their status in their communities as well as in their incomes and standards of living.

An organized group of these displaced home workers came together in 1811 under the leadership of "Ned Ludd." (There is some question as to whether Ned Ludd was a real person.) His followers, thus called the Luddites, began entering British clothing factories and smashing the stocking frames. They also sent letters to factory owners, threatening their lives.

By 1812 food prices were rising. The former cottage workers could no longer afford to feed their families, and food riots ensued. One such riot occurred at

Burton's Mill in Middleton. The owner of the mill, who knew that his use of machine knitting had angered hand workers in the local area, had hired armed guards to protect his factory. When the Luddites attacked, the guards killed three members of the group.

The local constabularies couldn't handle the Luddite attacks and therefore the British government became involved in the attempt to control the increasing level of violence. "Machine breaking," as the crime of destroying the stocking frames was called, became a capital crime in 1813. Although it's not clear from sources exactly how many were executed, the number appears to be between 60 and 70. Other individuals were punished by transportation to Australia.[1]

The Luddites had a surprising advocate: Lord Byron. His speech in the House of Lords on February 27, 1812 sympathized with the plight of the displaced workers and stated that "…the perseverance of these miserable men in their proceedings, tends to prove that nothing but absolute want could have driven a large, and once honest and industrious, body of the people, into the commission of excesses so hazardous to themselves, their families, and the community." Rather than turning the Luddites into criminals, he believed that "…had the grievances of these men and their masters (for they also had their grievances) been fairly weighed and justly examined, I do think that means might have been devised to restore these workmen to their avocations, and tranquillity to the country." Byron's appeal had little apparent effect because the "Frame Breaking Act" was passed over his objections.

In 1812, Luddite activities resulted in the death of a factory owner. The three men responsible were among those executed for machine breaking. After that point, the Luddites became less and less active and by 1817 were effectively nonexistent.

Depending on which historian's work you read, you may see the Luddites as a violent group resisting technological change or as a group opposed to the fixed prices imposed along with the factory-made goods. The latter point of view sees the Luddites as advocates for a free market economy rather than a violent rabble.

Nonetheless, it is apparent that the Luddites were not opposed to all technology, just the technology that was stealing their livelihoods. In that context, their actions become understandable, although not necessarily defensible.

AMISH

Probably the most well-known group of technology resisters in the United States is the Amish. As with the Luddites, however, the situation isn't as simple as it might first appear: The Amish attitudes toward technology are far more complicated than a simple refusal to use any technology at all.

---

[1] "Transportation" was a British punishment for those convicted of many types of crimes. It involved sending the prisoners to British colonies where they worked off their sentences as virtual slaves to wealthy British colonists. When their sentences were complete, most stayed in the colonies, although they were technically allowed to return to Britain.

Amish communities are first and foremost religious communities. They are named after Jacob Amman, a Mennonite leader who believed the Mennonites were becoming lax in their following of the dictates of their Anabaptist Christian religion. The Amish split from the Mennonites in 1693 and first came to the United States in the 18th century.

The Amish first settled on farms near what is now Lancaster, Pennsylvania. Today, Old Order Amish (those that follow the traditional Amish way of life) can be found in 21 states. There are about 55,000 living in Ohio, with about 39,000 in Pennsylvania and 37,000 in Indiana. The total population is estimated to be somewhere around 200,000.

Each Amish community follows a set of religious rules known as the Ordnung. Although there are some differences between communities, most adhere to the following:

- The Amish should live apart from the rest of society. They should not serve in the military nor do they accept any form of government assistance. This tenet is at the root of their choice not to use a great deal of technology.
- Amish life should be "plain." No one should have possessions that might make one person believe that he or she is better than another. The Amish reject pride and embrace humility. To this end, the Amish wear very plain, conservative clothing. Clothing does not use buttons or zippers but instead closes with hidden hooks and eyes.
- Children are not able to make the decision to be baptized into the church. Therefore, baptism occurs when people reach young adulthood.

The stricture to live apart from mainstream society is behind much of their decision to avoid technological devices. Instead of cars, they use horse-drawn carriages. Bicycles are also considered plain conveyances, and it's not unusual for young Amish to ride them.

Traditionally, the Amish do not use electricity. Kerosene lanterns and propane refrigerators are acceptable because they don't take power from the power grid. However, some Amish operate dairy farms and local health regulations require that milk be kept cool in an electrically powered device. The Amish keep themselves separate by generating the electricity for the machinery themselves, once again refusing to take it from the power grid. In addition, some Amish are embracing solar energy because it allows them to generate electricity and yet still remain apart from wider society. The electricity is acceptable if it is used primarily for business and home use is kept to a minimum.

The Amish attitude toward telephones follows the same thinking. Phones are necessary for business use, but it should be neither easy nor comfortable to use a phone.[2] Therefore several families share a phone that is installed outside, in a small shed. (Reportedly, some phones are even in outhouses.)

---

[2] One of the ironies of the Amish attitude toward telephones is that for some communities, cell phones are acceptable.

Amish do, however, accept modern medicine and its machinery for emergencies and life-threatening illnesses. They also ride in cars, trains, and planes when necessary to travel distances that aren't practical by horse and buggy. They run businesses that trade with those outside their communities. In other words, Amish isolation isn't total.

Sometimes an Amish community adapts a technology so that it falls within their guidelines. This is particularly true for farm equipment. Modern farm implements are acceptable if they are guided by a human or pulled by horse. It would therefore not be unusual to see a modern seeder pulled by a horse rather than a tractor.

When an Amish community member wants to adopt a new technology, the technology is brought before a group of community elders, who decide whether the technology is acceptable. Some items, such as the paper clip, are accepted without hesitation. Others, such as telephones, take considerable discussion and end with severe restrictions being placed on their use. In some cases, a technology may be rejected entirely, such as television.

The Amish therefore are not strictly technology resisters. They wish to remain apart from general society by maintaining their simple way of life, which rejects technologies that foster dependence on the outside world. Technologies that meet their criteria, or that they must use to do business with the outside, can be adopted or adapted. Others are rejected.

GREENPEACE

Some of the groups that resist technology do so because they believe technology has an environmental impact. One of the most well known is Greenpeace.[3] Its focus is on environmental damage caused by industry and technology. For example, its members work against commercial fishing because "industrial fishing" has reduced the population of large ocean-going fish. A typical action Greenpeace might take is to block the unloading of fish caught in violation of what Greenpeace believes was environmentally sound fishing policy. Greenpeace is also concerned with human contributions to global warming, toxic chemicals released into the environment, nuclear power, and bioengineering of food crops.

Greenpeace does not oppose all technology, only technology that has a negative impact on the environment: It is primarily an environmental group rather than an antitechnology group. To that end, Greenpeace publishes a "Guide to Greener Electronics,"[4] a ranking of technology companies as to their environmental friendliness. As of July 2007, Nokia was ranked as the most "green" technology company because it had completely eliminated polyvinyl chlorides (PVCs) from its manufacturing processes. Dell and Lenovo (the Chinese company that bought IBM's personal computer division) were tied for second. Apple, which ranked last in the previous edition of the Guide, moved up to 10th place because its CEO (Steve Jobs) promised to remove PVC from its products. The Greenpeace

---

[3] You can find Greenpeace's Web site at http://www.greenpeace.org/usa/
[4] The complete article about the Guide can be found at
http://www.greenpeace.org/usa/news/apple-greener-nokia-regains-l

campaign against Apple—called "Green my apple"—was waged worldwide, including print ads and in-person protest rallies. Greenpeace is also against the use of nuclear power for any purpose.

Much of what Greenpeace does takes place on the water. Beginning in 1971, members continue to confront those engaged in activities that Greenpeace believes harmful to the environment, often by blockading water-going traffic, even when the confrontation is significantly one-sided.

Greenpeace's efforts have been successful in many areas. It has stopped the killing of baby seals for fur. (Its protest technique was to spray paint red "Xs" on the baby seals' heads, making the white fur unmarketable.) Greenpeace also was successful in convincing the Shell Oil Company not to dispose of its offshore installation (the "Brent Spar") into the ocean.

Greenpeace has also had a major impact on worldwide agreements to limit whaling. Before the development of the whaling factory ships—and in particular, the development of refrigeration—whaling ships could only handle one whale at a time. Once the whale was rendered, the ship had to return to port before the meat would spoil. But the technology of t he factory ships makes it possible for whalers to stay at sea a long time, killing many whales and processing them quickly before returning to port. The result was the depletion of many whale species, which Greenpeace has worked to prevent. Greenpeace actions against whaling include protest demonstrations and actual interception of whaling factory ships at sea.

Greenpeace professes a creed of nonviolence. However, given the nature of its confrontations with those who are, in its members' opinions, harming the environment, the actions have occasionally become violent. For example, in 1994 Greenpeace members boarded a whaling ship and attempted to remove the ship's harpoon gun; the crew of the ship resisted physically to what they saw as an attempt to steal their property. Greenpeace members have also physically blocked access to PVC factories and stopped transport vehicles from entering or leaving.

Greenpeace has also been the victim of violence. In 1985, for example, two bombs exploded aboard the Rainbow Warrior, which was docked in France, killing a Greenpeace staff member. Greenpeace had been protesting French nuclear testing, and the bombing brought the organization a great deal of publicity.

The important thing to understand about Greenpeace is that it is not opposed to all technology but to the results of some specific technologies. This makes Greenpeace similar to both the historical Luddites, who acted against automated stocking frames, and the Amish, who reject only those technologies that are not consistent with their lifestyle.

NEO-LUDDITES

Although the groups about which you have been reading are not opposed to *all* technology, some groups and individuals would like to see the end of technology. People who think this way are often called *Neo-Luddites*. Their opposition to technology stems from the impact technology has had on the human condition, in particular, that it degrades humans and destroys many of the best features of our cultures.

The Neo-Luddites are a very loose group and often do not work together. Politically, they are both liberal and conservative. What brings them together is the belief that humans receive nothing positive from technology and that we ought to return to a time when we were less dependent on it. Their objection to technology might therefore be considered to be more philosophical than anything else.

One of the most well known of the Neo-Luddites is Kirkpatrick Sale. He advocates the destruction of technology (property, not people) and finds few technologies to be positive. Consider, for example, his response to an interviewer—who described the car as an "emancipating" technology—when asked about the automobile:

> Only someone ignorant of industrialism and the Enlightenment mind-set would have thought the automobile "emancipating." It was intended to increase consumerism, individualism, anomie, community disintegration, and the power of markets, and it did. Similarly with most current technologies. Once we understand that technologies are not either accidental or neutral we will understand that they inevitably express the values and beliefs of the powers in society that introduce and adopt them; a progressive nation-state capitalism will produce one kind of technology, a decentralized tribal anarchocommunalism of an entirely different kind.[5]

Sale believes that humanity has a very short life span (another 25 years) and that even if we manage to survive longer, we will have learned that the negative aspects of technology outweigh any benefits. He believes that we are heading for a "technological Armageddon." Revolt is useless: "The nation-state, synergistically intertwined with industrialism, will always come to its aid and defense, making revolt futile and reform ineffectual."[6] Nonetheless "...resistance to the industrial system, based on some grasp of moral principles and rooted in some sense of moral revulsion, is not only possible but necessary."[7]

Other well-known Neo-Luddites include Jerry Mander ("Jerry's message: Technological civilization is destroying nature and human life. Jerry's solution: Dismantle technological civilization. Simple as that."[8]) and Steve Talbott, who believes that technology has interfered with education, that technology has had a negative effect on the environment, that computers have eroded language, and that "computer-based organizations...sustain themselves in a semi-somnambulistic manner, free of conscious, present control."[9]

Neo-Luddites are by and large nonviolent. They write books and articles and make speeches to interested audiences.[10] They hope to influence others, resulting in a groundswell of popular opinion against the use of technology.

---

[5] From http://www.primitivism.com/sale.htm
[6] Originally published in The Nation, June 5, 1985 but now available at
http://www.geomatics.ucalgary.ca/~terry/luddite/sale.html
[7] Also from Sale's article in *The Nation*.
[8] From http://www.undueinfluence.com/mander.htm. For more on Mander's philosophy,
see http://www.ratical.org/ratville/AoS/theSun.html#IV
[9] From http://natureinstitute.org/tech/index.htm
[10] Kirkpatrick Sale was once given four minutes to speak. He spoke for one and a-half minutes and took up the rest of the time using a sledgehammer to smash a computer. He said that smashing the computer was quite satisfying, but he would never do anything similar to a person. He is in favor of violence to property but not to living beings.

ANARCHO-PRIMITIVISM: THE RETURN-TO-NATURE MOVEMENT

Anarcho-primitivists (or just primitivists, for the purposes of this book) are those who believe we should return to a hunter-gatherer type of life, even forgoing agriculture. They are anarchists, believing there should be no governments at all, much less any technology.

Primitivists believe that the first step down the path toward the problems we have today began when humans began to farm rather than gathering foods growing wild in nature. Farming made humans dependent on technology, even if it was something as simple as a hoe or a horse-drawn plow. Simple tools—something a person picks up from nature, uses for a while, and then discards—are acceptable, but not permanent tools that are manufactured in any way.

They look at archeological evidence that indicates the early hunter-gatherer communities were peaceful and see them as a model for humanity. In their opinion, the inhabitants of primitive communities had a great deal of leisure time, were free from "organized violence," were healthy, lived in harmony with the natural world, and had gender equity.

To achieve a hunter-gatherer lifestyle, primitivists believe we should disassemble all technology along with all other trappings of civilization. They are also concerned that language, writing, and other physical forms of representing information ("symbolic culture") constrain human understanding of the world and therefore advocate eliminating them as well. Primitivists are against what they call "domestication," anything that takes humans away from a primitive, nomadic lifestyle. In addition, they reject scientific discoveries.

One of the ironies of the primitivist movement is that because primitivists are against organizations, they are also against any type of organized revolt against civilization. Nonetheless, they advocate revolution rather than reform (the latter being modification rather than dismantling of something). Primitivists are by and large nonviolent. Although they have no formal associations, several writers (for example, John Zerzan and Daniel Quinn) are seen as the voice of the movement. In addition, primitivist thought is espoused on the Web and in magazines (for example, *Green Anarchy*).

WHEN RESISTANCE TURNS VIOLENT: THE UNABOMBER

For the most part, modern resistance to technology has been nonviolent. Occasionally, however, an individual takes that resistance too far. Theodore (Ted) Kaczynski, for example, killed and injured a number of people in his attempts to destroy and discredit technological advancement.

Kaczynski's story begins normally enough: He was a bright student who earned a PhD in mathematics from the University of Michigan. He then began a fairly standard academic career by becoming a faculty member at UC Berkeley. Something happened, however, while he was at UC Berkeley, although no one knows exactly what triggered the change. Kaczynski gave up his academic position and went to live a primitivist lifestyle in the Montana mountains.

Kaczynski began his actions against technology in 1978 by mailing a package bomb to Northwestern University; two people were injured when the bomb exploded. Over the next 17 years he sent increasingly dangerous package bombs to people associated with science and technology, primarily at academic institutions. His targets also included an American Airlines flight; the bomb was triggered by an altimeter.[11]

Kaczynski's package bombs killed three people: a computer store owner in 1985, an advertising executive whose company did work for Exxon in 1994, and a timber industry lobbying group's president in 1995. Kaczynski, who was nicknamed the Unabomber by the media, was wanted by the FBI for capital crimes.

Kaczynski picked his targets because they were in some way associated with what he considered to be the worst side of technology. In 1995 he came out of the shadows and mailed letters to some of his victims explaining what he was trying to do. He also indicated that if his "Manifesto" were published in a major newspaper, he would stop the bombing. The newspapers initially weren't sure whether publishing the 35,000 word document was appropriate; they hesitated to give a murderer a platform. But given that publication would stop the bombing, the *New York Times* and the *Washington Post* decided to go ahead with publication. Kaczynski's manifesto appeared on September 19, 1995 in its entirety.

The Manifesto begins with the following:

> The Industrial Revolution and its consequences have been a disaster for the human race. They have greatly increased the life-expectancy of those of us who live in "advanced" countries, but they have destabilized society, have made life unfulfilling, have subjected human beings to indignities, have led to widespread psychological suffering (in the Third World to physical suffering as well) and have inflicted severe damage on the natural world. The continued development of technology will worsen the situation. It will certainly subject human beings to greater indignities and inflict greater damage on the natural world, it will probably lead to greater social disruption and psychological suffering, and it may lead to increased physical suffering even in "advanced" countries.[12]

Like other primitivists, Kaczynski wanted to undo all the changes that technology had made to society. The major difference between he and others who espouse the same philosophy is that most primitivists don't resort to murder.

Kaczynski was caught when his younger brother, David, suspected that the writing style of the Manifesto matched that of his brother. The younger Kaczynski turned over some handwritten letters from his brother to an FBI profiler, who concluded that the letters and the Manifesto were written by the same person. Based on that evidence, the FBI issued an arrest warrant.

---

[11]Airlines were not always as concerned about security as they are today. Security screenings didn't even begin until 1972, after there had been several plane hijackings. Initially, security measures consisted of simply walking through a metal detector. It wasn't until 1988 and the aftermath of the bombing of Pan Am flight 103 over Lockerbie, Scotland that airlines began to screen carry-on bags. Therefore it was easy for Kaczynski to smuggle a bomb on board a plane in 1978.

[12]For the complete text of Kaczynski's published manifesto, see http://cyber.eserver.org/unabom.txt

Kaczynski was arrested at his cabin in the woods on April 5, 1996, where the FBI found bomb-making materials. Although a court-appointed psychiatrist stated that Kaczynski was suffering from paranoid schizophrenia, he was considered competent to stand trial. He avoided the death penalty by pleading guilty to all the charges against him and to this day remains in prison in Florence, Colorado, serving a life sentence without possibility of parole.

IRONY OF TECHNOLOGICAL RESISTANCE: USING THE INTERNET TO SPREAD THE WORD

There is a funny bit of irony surrounding resistance to technology: Most of the organized resistance groups use the Internet. They use the technology they are supposedly resisting to spread the word about their philosophies. Even the primitivist magazines have Web sites.

Although the Amish do not use the Internet directly for personal or business purposes, they are well aware of the economic benefits of selling their goods over the Web. They therefore have non-Amish associates establish Web sites to sell Amish goods—quilts, furniture, and so on—on their behalf.[13] There are also educational Web sites, operated by non-Amish, to which the Amish provide information.[14]

---

[13] You can find a number of links to businesses that sell Amish goods at http://www-personal.umich.edu/~bpl/mennocon.html#amish. Scroll down until you see the section labeled "Amish Connection."

[14] As an example of an educational Web site maintained by non-Amish people on behalf of the Amish, see http://holycrosslivonia.org/amish/

WHERE WE'VE BEEN

Although most people in Western societies welcome most technological advances, there have been (and are) groups and individuals who have resisted the introduction of new technologies. The most well-known historical group is the Luddites, which arose in early 19th century England to protest the loss of jobs and product quality that were the result of the introduction of automated knitting machines.

Today, the Amish are among the most well-known groups of technology resisters. A religious community, the Amish choose to live apart from mainstream society. They therefore reject technologies, such as electricity from the community power grid, that would connect them to the outside world. They do not reject all technology but restrict its use to those things that are consistent with their simple way of life.

Environmental groups, such as Greenpeace, also reject some forms of technology. In this case, they reject anything they perceive as having a negative impact on the natural world.

Neo-Luddites view technology as dehumanizing and believe we should significantly reduce our use of it. This philosophy is carried to its extreme by the anarcho-primitivists, who espouse dismantling all of civilization and returning to a hunter-gatherer lifestyle.

For the most part, modern resisters of technology are nonviolent. However, individuals such as Ted Kaczynski have turned to violence (including murder) in protest of what they believe to be the evils of technology.

THINGS TO THINK ABOUT

1. Find a classmate, friend, or relative who doesn't like to use a computer. Why is that person resistant to computer use? Is there anything that someone could do to get the person to use a computer? If so, what?

2. There has been a great deal of research into the human resistance to change. Use the Web to find some of that research and identify things that organizations can do to make it easier for people to accept change.

3. Imagine you were a hand knitter who made stockings. Short of breaking the stocking frames that were turning out cheaper stockings, what could you have done to preserve your income, your status in the community, and the quality of the goods?

4. Read the entire Unabomber's Manifesto at http://cyber.eserver.org/unabom.txt. List the author's arguments against technology; list his potential solutions to the problems he perceives. For each item in your list, indicate whether you think his point is valid and give a reason for your decision.

5. Many groups that resist technology have resorted to violence. Some might argue that such violence is born of frustration with government agencies and other organizations that won't listen to them otherwise. Do you believe that violence is ever justified in the resistance to a technology? If so, under what circumstances would you consider violence as a reasonable action? If not, why not.

WHERE TO GO FROM HERE

Luddites and Neo-Luddites

Jones, Steven E. *Against Technology: From Luddites to Neo-Luddism*. Florence, KY: Routledge, 2006.

Pynchon, Thomas R. "Is it O.K. to be a Luddite?" *http://www.themodernword.com/pynchon/pynchon_essays_luddite.html*

Sale, Kirkpatrick. *Rebels Against the Future: The Luddites and Their War on the Industrial Revolution*. New York, NY: Basic Books, 1996.

Amish

Legal Affairs. "The gentle people." *http://www.legalaffairs.org/issues/January-February-2005/feature_labi_janfeb05.msp*

Levinson, Paul. "The Amish get wired. The Amish?" *http://www.wired.com/wired/archive/1.06/1.6_amish.html*

Powell, Kimberly, and Albrecht, Powell. "Amish 101-Amish beliefs, culture & lifestyle." *http://pittsburgh.about.com/cs/pennsylvania/a/amish.htm*

"The Amish: simply captivating." *http://www.padutchcountry.com/our_world/the_amish.asp*

Anarcho-Primitivism

Beating Hearts Press. "Anarchist distribution for Australasia." *http://www.beatingheartspress.com/anticiv.html*

"Green anarchy: an anti-civilization journal of theory and action." *http://www.greenanarchy.org/*

Green Anarchy Collective. "An introduction to anti-civilization anarchist thought and practice." *http://www.greenanarchy.org/index.php?action=viewwritingdetail&writingId=283*

Zerzam, John. *Against Civilization: Readings and Reflections*. Los Angeles, CA: Feral House, 2005.

The Unabomber

CNN Interactive/Time. "The unabomb case." *http://www.cnn.com/SPECIALS/1997/unabomb/*

Court TV News. "Psychological evaluation of Theodore Kaczynski." *http://www.courttv.com/trials/unabomber/documents/psychological.html*

News Real. "Revolutionary suicide." *http://www.salon.com/news/1998/01/21news.html*

Accessibility of Technology 5

WHERE WE'RE GOING

In this chapter we will

- Discuss a variety of profiles that describe Internet users, including where they live and demographics such as income, education, and gender.
- Consider the barriers that exist to prevent people from gaining access to technology.
- Examine the "digital divide" to determine its extent.
- Look at proposed solutions to the digital divide.

INTRODUCTION

Any person reading this book probably has easy access to technology. But that is not true of everyone in the world. In this chapter we will look at how technology has spread across the globe and focus specifically on who has it and who doesn't.

WHO AND WHERE ARE THE USERS?

One of the events that technology futurists were unable to predict was the expansive growth of the Internet. Each year millions of people who weren't connected before gain access. Although our definition of technology extends far beyond computers and the Internet, the most reliable technology-use statistics related to the Internet, so we will use those numbers to explore the dispersion of technology. The statistics presented in this section were current when this book was written; today, the numbers will certainly be greater!

Geographical Differences

Figure 5.1 shows the distribution of Internet users by region throughout the world.[1] Those of us in North America tend to believe we are the biggest users of the Internet, but there are more users in both Asia and Europe. However, the raw number of users really doesn't give you a good picture of what is happening.

More telling is what we call "Internet penetration," the percentage of a population that has access to the Internet. The access can be at home, on the job, or through a public venue, such as a library or Internet cafe. When we look at Internet penetration (**Figure 5.2**), we can see that North America, Australia, and Europe have the highest percentages of connected people. Asia may have more users, but given Asia's high population, that raw number of users translates into a very small penetration.

How fast is Internet access growing? During the years 2000 to 2007, African use grew the most (**Figure 5.3**), but considering their small penetration rate, strong growth isn't surprising. North America had the lowest growth. That shouldn't be surprising either, because North Americans tend to be early technology adopters and many North American users gained their Internet access before 2000.

We can also look at Internet statistics in terms of which countries have the highest percentage of the world's Internet users. The darker gray bars in **Figure 5.4** represent the percentage of world users found in the top 10 countries in the world. Although China has a much larger population than the United States, the United States still has a much larger proportion of users. The lighter gray bars in Figure 5.4 represent the percentage of users who have broadband connections to the Internet. Just under 20% of U.S. users have broadband access, but more than 27% of South Koreans have high-speed access. France, which was a leader in deploying high-technology communications to its citizens in the 1980s, also has high broadband connectivity.[2]

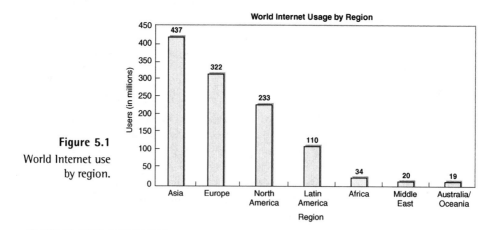

Figure 5.1
World Internet use by region.

---

[1]The statistics in this section come from http://www.internetworldstats.com/stats.htm, which, as far as I can tell, has the most up-to-date Internet usage statistics available for free, anywhere.

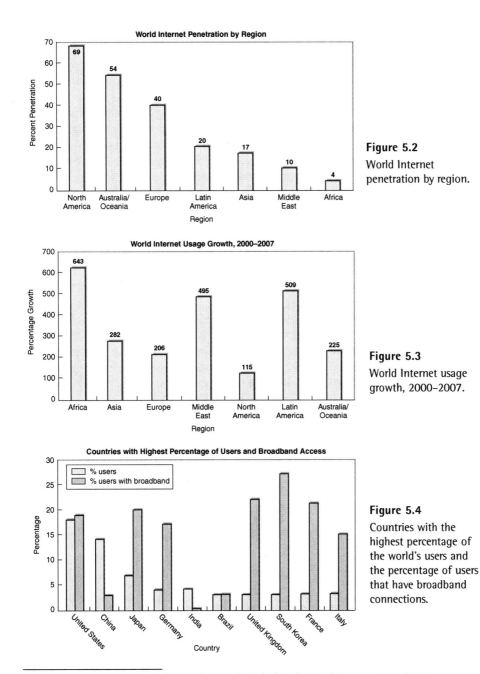

Figure 5.2

World Internet penetration by region.

Figure 5.3

World Internet usage growth, 2000–2007.

Figure 5.4

Countries with the highest percentage of the world's users and the percentage of users that have broadband connections.

---

[2] In the early 1980s, France replaced many of its standard telephone lines with ISDN (Integrated Services Digital Network) connections. ISDN, which is largely outdated today, is a dial-up digital service that is faster than using a modem for data communications. The French Minitel system, which used the ISDN lines, included an electronic phone book, electronic greeting cards, and e-mail. French users were given Minitel terminals for free and then paid for the time they used the system. For information about the project, see http://en.wikipedia.org/wiki/Minitel

Demographic Differences

The statistics presented in the preceding section don't reveal how Internet access is distributed throughout a population. In the United States, for example, there are differences in Internet use among age groups, economic groups, and education level.[3]

Throughout human history, younger members of society have been more open to change and new ideas than their elders. It should therefore come as no surprise that the adoption of computing technology follows a similar pattern. Although Internet use rises across all age groups from 1995 to 2005, the younger the age group, the higher the percentage of individuals who are online (**Figure 5.5**).[4]

Internet connectivity is positively correlated with income level.[5] **Figure 5.6** shows that more than 90% of users in the highest income group are connected but only 48% of those in the lowest income group have access. Given that computers and monthly Internet service provider (ISP) charges are beyond the means of the poor and many of the working poor, this result isn't a surprise.

Internet access is also positively correlated with educational level (**Figure 5.7**). The percentage of individuals with Internet access rises with the highest level of education. This, too, isn't a surprising statistic, because higher levels of education have traditionally been associated with a curiosity about new technologies and a willingness to learn about them.

The preceding statistics don't tell the whole story, however. There are also gender differences in Internet use that relate to *what* people do rather than *how many* are online. By 2005 the percentage of American men and women using the

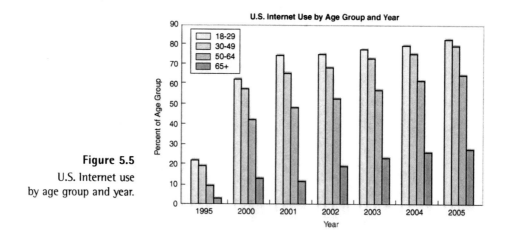

Figure 5.5

U.S. Internet use by age group and year.

---

[3]The data on which the charts in this section are based come from http://www.cnn.com/SPECIALS/2005/online.evolution/interactive/chart.online.life/intro.html

[4]There are exceptions to every pattern, such as the decreasing use of the Internet associated with increasing age. Just about everyone knows a senior citizen who is a computer whiz, uses e-mail regularly, and surfs the Web with ease. Nonetheless, seniors are much less likely to be online than those in younger age groups. Whether this pattern remains as those who have grown up with computers age is open to question. I would predict that it will not, because we tend to remain comfortable throughout our lives with those things that we were comfortable with growing up.

[5]These numbers are from 2004.

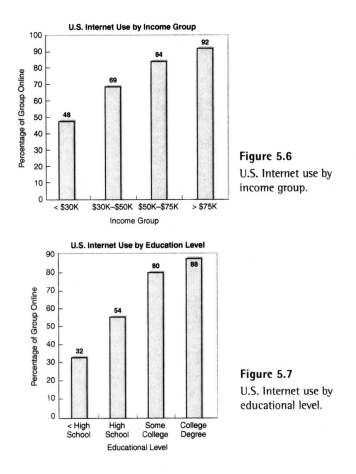

Figure 5.6
U.S. Internet use by income group.

Figure 5.7
U.S. Internet use by educational level.

Internet was about equal. A Pew Internet and American Life survey[6] revealed that 68% of American men and 66% of American women used the Internet. However, there are more women than men online because women make up more than half the population.

The age distribution between American men and women online isn't equal. At younger ages women are more likely to use the Internet, whereas in the senior years men are the heavier users. There are also racial differences (**Figure 5.8**). Notice that more black women use the Internet but more men in the "Other" category (which includes Asians) are online.

Married Americans are also more likely to use the Internet as opposed to unmarried individuals. Parents of children under 18 are significantly more likely to use the Internet as those without young children.[7]

The Pew survey also looked at how much time men and women spend online. Men spend more time connected and go online more often than women,

---

[6]http://www.netlingo.com/more/PIP_Women_and_Men_online.pdf
[7]This result isn't surprising, given that young children increasingly use the Internet for school-related research and do so with the help of their parents.

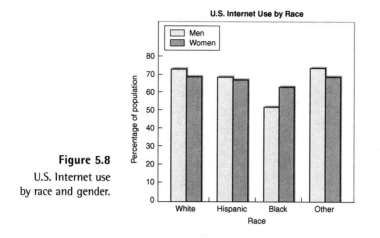

Figure 5.8

U.S. Internet use
by race and gender.

although both genders have equal access to the Internet at home. Internet use at
use at work is relatively equal for men and women.

According to the Pew survey, women engage in the following online activities more often than men:

- Sending e-mail
- Getting maps or directions
- Searching for health and medical information
- Finding help for physical and psychological problems
- Finding religious and spiritual information

The Pew survey indicates that the percentage of men and women who
purchase online is different. However, anecdotal evidence and data collected by
my students contradict that finding and suggest that women are far more likely to
purchase items online than men. This is as expected, given that women do more
of a family's shopping in stores than men.

In contrast, men are more likely to do the following:

- Research (both products and services and work related)
- Check weather
- Get news (including sports news)
- Find do-it-yourself directions
- Download software, music, and videos
- Get financial information, manage investments, and pay bills
- Contribute to chat rooms and blogs
- Take an online class

Men are also much more likely to visit adult Web sites. In fact, surveys
conducted by my students indicate that college-aged women are highly unlikely
to visit adult sites, whereas a small percentage of men do so. (In the Pew survey,
21% of the men versus 5% of the women accessed the adult sites.)

Both men and women use the Internet as a communications tool. However,
a higher percentage of women send e-mail, use instant messages, send electronic

greeting cards, and exchange text messages than men. Men are more likely to participate in chat rooms and discussion groups as well as use Internet telephony.

Men seem to be more knowledgeable about computing and technology in general; they are also more likely to be responsible for computer maintenance. In addition, men are more likely to try new technologies. However, a 2004 Pew survey found that teenaged women are just as likely as their male peers to try new technologies and that their knowledge of technology rivals that of teenaged men.

BARRIERS TO ACCESSIBILITY

What accounts for the differences in world Internet access? There are two primary barriers: economics and politics. In many cases, the two barriers work together to prevent people from gaining access. There are also reasons that people who can afford computers and who aren't restricted by political actions do not access the Internet. We will look at those reasons at the end of this section.

Economic Barriers

Although you can surf the Web for free, the Internet isn't precisely free. At the very least, most people (or their employers) pay something just to connect. Some free access is available through public libraries, but beyond that an individual or company is paying for the computer and the connection through an ISP. Internet cafes, which are popular throughout the world, charge by the minute or hour. And if you consider that taxes pay to support libraries, then library access isn't totally free either.

The upshot is that if you can't afford a computer and an ISP account, can't afford an Internet café, and don't feel comfortable going to the library to use a computer, then you're cut off from the Internet.

Political Barriers

Some people are unable or unlikely to connect to the Internet because their countries are in political turmoil. A refugee in Darfur, for example, is probably more worried about survival than about technological access. But beyond war-torn areas, there are countries that actively regulate what their citizens can access over the Internet. Cuba and China are the most well known, but some countries in the Middle East also restrict access.

Most democracies believe that an educated public is essential if the citizens are to make good decisions about how their country is to be governed. One way that totalitarian governments maintain their control over their populace is to restrict access to information. Cuba is one of the few Communist governments left in the world; China, although it has moved to a free market economy, is still a totalitarian state that restricts freedom of expression.

Cuba As of June 2007 only 1.7% of Cuba's population had access to the Internet. A government edict issued in January 2004 made it illegal to use Cuba's normal telephone network to dial up ISPs. Cubans can go to the post office to send and receive e-mail and to access the Cuban intranet (populated by Cuban Web sites) but not the Internet. Even at that, access isn't cheap: Three hours costs about $4.50, roughly one-third of the average Cuban's monthly salary. However, many Cubans have access to e-mail and the Cuban intranet in their workplaces.

Those who want access to the Internet from their homes must use an alternate telephone network for their data calls. Time on this network must be paid in U.S. dollars at the rate of 8 cents per minute, a cost far beyond the means of most Cubans. The result is that access to the Internet is restricted to only the wealthiest Cubans, a group that is overwhelmingly in support of the current Cuban government.

The Cuban government has stated publicly that the reason for the restriction is limited bandwidth, that restricting data use of the telephone system is a fair way to allocate the use of a limited resource. But groups opposed to the Cuban government believe that Internet access is given only to those who support the current government and that the Internet restrictions are one way of shutting down dissent.

China The Chinese government is much more overt than the Cuban government in its efforts to restrict access to information. Although Chinese citizens can access the Internet, the Web sites they can reach are limited by filtering software run by ISPs. In 2002, two researchers (Zittrain and Edelman) found that of 204,012 Web sites they visited, more than 50,000 could not be reached from inside China. Blocked sites included major news networks (for example, ABC and CBS), major universities (for example, MIT, Columbia, George Washington University), research centers, non-Chinese government agencies, religious organizations (including the banned Falun Gong), sexually explicit sites, and all political sites that do not support the current Chinese government.

The filtering is very effective. In a 2004 update to their original study, Zittrain and Edelman call it "persuasive, sophisticated, and effective."[8] The Chinese government has also requested that major search engines, including Google, filter the results of searches that originate in China.

Google agreed to China's requests to censor its search results and now filters search results that are sent to China. Although Google states that it isn't in favor of censorship, if it wouldn't do the filtering, then it wouldn't be able to reach China at all.[9] Other search engines also censor search results destined for China.

China isn't the only country that filters Internet access by its citizens. Bahrain filters a few sites (some pornography, religious, and political sites); Burma and Iran are as restrictive as China, if not more so. The list of nations that perform Internet filtering also includes Saudi Arabia, Singapore, Tunisia, and the United Arab Emirates.

Other Reasons for Not Using the Internet

Even if there aren't any economic or political barriers, some people simply aren't connected to the Internet. Most of these individuals have made the choice not to connect for one or more of the following reasons:

---

[8] http://www.opennetinitiative.net/studies/china/
[9] For details, see http://news.bbc.co.uk/2/hi/technology/4647398.stm

- Technology is a bad thing (for example, the Neo-Luddites).
- Computers are just too frightening.
- Computers are too complicated to learn to use.
- There is nothing of value on the Internet.
- There is too much danger from pornography, sexual predators, and so on in the online world.
- The Internet is not secure (for example, identify theft).

Some of these nonusers have never used the Internet. Others have used it for a while and decided not to use it any more. And still others may not use the Internet directly but have others do it for them. Regardless of the reason, however, the people who choose to avoid using computers and the Internet seem unlikely to change their attitudes without the help of people in favor of technology access.

THE GREAT DIGITAL DIVIDE

The term "digital divide" refers to the gap between those who have access to computing technology (and, in particular, the Internet) and those who don't. There is certainly a digital divide between those who choose to avoid technology and those who choose to use it (assuming there are no other barriers to use). This digital divide is of their own making, but what about those who want computing and the Internet but face a major barrier to doing so?

Is There a Digital Divide?

The statistics presented earlier in this chapter make it clear that not everyone in every country has access to the Internet, and the barriers to access suggest that not everyone who lacks access is choosing to avoid the Internet. This strongly implies that the digital divide is real, that there are those who are being left out of technological advances, many of whom haven't even been given the opportunity to make a choice. Nonetheless, researchers differ in their opinion as to whether this divide is permanent, growing, or important.

One group of researchers, led by the work of Everett Rogers, points to technologies such as television and telephones as an example. When those technologies were new, there was a significant gap between those who had them and those who didn't. In the case of the telephone, the gap between those who had one and those who didn't was based both on economics and on whether telephone lines were available. In the case of televisions, the gap was primarily economic, although there were (and still are) areas of North America that couldn't receive over-the-air broadcasts.[10] However, as the continent became laced with telephone lines and the cost of telephone service and televisions declined, the gap between those who had the technologies and those who didn't was virtually eliminated. By 1978, for example, 98% of U.S. homes had at least one television.[11]

---

[10]Don't believe it? I live in just such a "white area," half way between New York City and Albany. We can't receive broadcasts from either major city, and our local stations are limited to one religious channel and a talk channel, which can be picked up only with a good roof antenna. Until cable came into the area, people used C-band satellite dishes (the 10-foot mesh dishes) to receive television; today we use either cable or digital satellite (the small solid dishes).

[11]For more statistics about television use and ownership, see http://www.tvhistory.tv/facts-stats.htm

From his research, Rogers was able to create the technology diffusion curves shown in **Figure 5.9**. The right graph plots the sales of a technology over time; when viewed as a cumulative plot, sales or penetration of technology forms an S-shaped curve. Rogers divided the bell-shaped curve into five segments, from left to right:[12]

1. 2.5%: innovators
2. 13.5%: early adopters
3. 34%: early majority
4. 34%: late majority
5. 16%: laggards

Researchers postulate that access to computing and the Internet will take the same path as the technologies Rogers investigated. Certainly prices of desktop computers have declined significantly, but even a $300 computer may be beyond the budget of lower income families. The ongoing cost of an Internet connection is also more than many of the poor can handle. As we saw in the first part of this chapter, Internet penetration has not come close to the 98% penetration of the television set. Technologies such as television and telephones also don't engender reasons for nonuse that are as strong as those against computing.

Nonetheless, those who believe that the digital divide will go away on its own point to the availability of computers in libraries and other public places; a person doesn't necessarily need to own a computer to have access to the Internet. This ignores, however, the inconvenience of having to go to a special location to use a computer. Libraries may not be located within walking distance and therefore require either a drive or a bus ride to reach them; some people may simply find it too much trouble to make the effort to reach a computer or they may be physically unable to make the trip. In addition, many libraries have a limited number of machines, for which there is often a waiting list. Users must reserve time on a computer in advance and usage time is limited, often dependent on the demand for the resource.

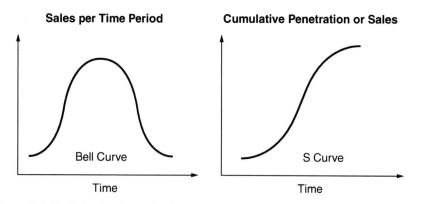

Figure 5.9 Typical technology adoption curves.

---

[12]For original reports on Rogers's research on the diffusion of technology, see http://www.12manage.com/methods_rogers_innovation_adoption_curve.html and http://www.kzero.co.uk/blog/?p=216

Disadvantages of No Technology Access

If you don't have access to current information technology, what do you lose? In the opinions of those who support the use of the Internet, you may lose any or all of the following:

- Access to employment opportunities that are posted exclusively on the Web.
- Ability to apply for jobs that require that resumes be e-mailed.
- Ability to have jobs that require technology skills.
- Access to up-to-date, in-depth news. Newspapers can be as much as 24 hours behind in their reporting; television news, which is the primary source of news for any people, is limited in depth because news shows are limited in time.
- Access to potential markets. Businesses of all sizes have seized the opportunity to expand their markets by selling over the Internet.
- Access to a wide range of products and merchandisers. If you don't shop on the Internet—and you don't shop by catalog using the telephone—you are limited to stores you can reach by foot, car, or bus. This restricts which products you can purchase as well as the stores from which you can purchase them. As discussed later in this book, shopping online doesn't necessarily get you a better price for an item, but it certainly opens up more opportunities for comparing product features and prices.
- Access to research information. Before the widespread growth of the Internet, a student who wanted to research newspaper, magazine, and journal articles used paper indexes such as *The Guide to Periodical Literature*, *Social Science Citation Index*, and *Chemical Abstracts*. Libraries provided these resources for their users; they were far too expensive (and took up too much shelf space) for an individual to own them. Once you found the articles you wanted, you would then search the library shelves for the appropriate publications. If the library didn't have what you needed, then you resorted to Interlibrary Loan and waited for the article to come from another library. Today, most research material is available over the Internet, although not all of it is free. Students turn to the Internet first and print materials second. At the same time, some scientific journals have turned to electronic-only publication because of the high cost of print publication.

Proposed Solutions to the Digital Divide

Giving Internet connectivity to those who don't have it has been seen as an answer to a number of problems. For example, a project to help the homeless find jobs involved giving them e-mail accounts and teaching them to use the Internet for job searching. There were two reasons behind the idea: Using the Internet would give the homeless greater access to job listings, and it would also hide the fact that they *were* homeless.[13] The homeless would then be on a par with others seeking

---

[13]The Internet has been called "the great leveler" because computers hide the physical circumstances of the user. A small business can look as professional as a large business, and age, gender, economic status, and appearance can be hidden easily.

jobs and have a better chance at finding employment that could give them the means to afford permanent housing.

Other projects have been more ambitious. In the rest of this section we will look at two specific projects that are attempting to bring economic benefits and educational benefits to countries without significant Internet penetration, discuss some additional projects, and whether any of these projects will work.

Cell Phone Project In 1997, Iqbal Quadir founded Grameen Phone, a company designed to bring wireless phone service to rural villages in Bangladesh. Quadir had three goals:

1. To allow rural villages to connect to the rest of the world
2. To provide business opportunities
3. To foster a sense of entrepreneurship that can translate into economic development

Currently, his company offers phone service to more than 80 poor villages. The company goes into the village and supplies at least one villager with a phone. The single person with the phone then charges villagers for use of the phone. The phone connects the villagers with the outside world, and the person managing the phone has a business that not only pays its own expenses but provides profit for the operator and demonstrates to the villagers that it is possible to run a profitable business in the village. The contact that the phone provides to the outside world has expanded the markets for the goods produced by village residents and has therefore contributed to the economic well-being of the village, beyond the business run by the phone operator.[14]

This type of project is a small step toward bridging the digital divide. It uses technology to attack the poverty in the villages and paves the way for more opportunities to use technology.

$100 Laptop Project The One Laptop Per Child (OLPC) project is an attempt to provide low-cost ($100 or less) computers for schoolchildren in the developing world.[15] Begun officially in 2005 by Nicholas Negroponte, the research that suggested the project was performed by Negroponte and Seymour Papert over the preceding 10 years. The basic idea is that by allowing children in developing countries to use computers in school and later taking those computers home for family use will provide the children with opportunities they would not have had otherwise: "Our goal: To provide children around the world with new opportunities to explore, experiment, and express themselves."[16]

The core of the project is a rugged laptop that not only costs just under $200 but has batteries that can be charged in a variety of ways, including human power.

---

[14] To view a speech by Quadir about his efforts, see http://www.ted.com/index.php/talks/view/id/79. While you're there, you can explore the TED Web site and view videos of other talks given by technology innovators. TED is a fascinating organization that brings together "forward thinkers" once a year to exchange ideas. Admission to its conference is by invitation only (assuming you can afford the $4,000 fee).
[15] The OLPC group's Web site can be found at http://laptop.org
[16] From http://www.laptop.org/en/vision/index.shtml

(Many of the places for which the laptop is designed have no electricity.) Known as the XO, the laptop (**Figure 5.10**) became available in 2008.

The XO has the following features, all of which are designed to make the machine low cost, functional, and able to tolerate harsh treatment:

- An AMD Geode LX-700 CPU.
- 256 Mb DRAM.
- 1 Gb flash storage. There is no hard drive because hard drives are too easily damaged.
- A 7.5" low-power LCD display.
- A complete wireless router. This allows children to connect to each other's machines as well as to the Internet.
- A touch pad pointing device. A mouse, which can't be integrated into a machine, is too easily lost.
- A battery that can be charged from a variety of sources, including wall power, a car battery, and human power with a crank, foot pedal, or pull cord.
- Open source, free software. The machine's environment is open so that those who have the desire and knowledge can either modify the machine's software or write their own programs. Included software is focused on activities rather than on applications. For example, one major activity is the Journal, through which students can record what happens in their everyday lives.

The OLPC project envisions governments purchasing large numbers of the XO and distributing them to children. Peru, Rwanda, Uruguay,[17] Libya, Brazil, Thailand, and Nigeria are among those countries committed to purchase machines. Libya, for example, has ordered 1.2 million units, one for every school child in the country.

Figure 5.10
The XO laptop.
(Courtesy of Mike McGregor.)

---

[17]See http://wiki.laptop.org/images/1/1a/San_Jose_Mercury_News_-_LAPTOPS_AND_LESSONS_REACH_LATIN_AMERICA.pdf

The OLPC project raises some interesting questions. First, will children use the computers in the way they are intended? Pilot projects suggest they will. In fact, when students take the computers home, often an entire family uses them, giving the project a wider impact than just school-aged children. Second, and perhaps more importantly, will this childhood experience translate into adult opportunities? Will it encourage young people to get further education, to start businesses in their villages, to engage in projects that improve life in their communities? Because the OLPC project hasn't had a wide deployment as of the time this book was written, it's too soon to tell. However, it will bear watching to see if this effort to bridge the digital divide will benefit the students and their families.

Other Projects to Bridge the Digital Divide There are a number of other projects underway in attempts to bridge the digital divide throughout the world, some under the aegis of nonprofit groups and others funded by governments:

- CNET Networks International Media and the UK charity Computer Aid International have formed Computer Aid (a nonprofit group), which takes donations of used computers, refurbishes them, and donates them to schools and community groups throughout the developing world. It has placed at least 42,000 computers in 90 countries.[18]
- Washington State University operates the Center to Bridge the Digital Divide. One of its projects (Last Mile Initiative Rwanda[19]) is working to develop elearning courses to support village Internet connectivity, with a focus on establishing Internet-based businesses. With help from USAID,[20] a U.S government agency, it has distributed wireless communications technology, including satellite long distance service. Interns from the university work with local residents to establish centralized communications facilities (telecenters) with Internet access at gathering places such as coffee cooperatives. In addition, the Center participates in Horizons,[21] a project aimed at eliminating poverty in small communities in the U.S. upper Midwest and Pacific northwest.
- The Community Technology Centers project,[22] administered by the U.S. Department of Education, establishes community-accessible computing facilities in low-income communities (urban and rural) throughout the United States. Its major focus is on preschool education, but it also supports educating adults in the uses of technology so they can obtain better jobs and provides after-school programs for older children and teens.

---

[18]For more information about Computer Aid, see http://www.bridgethedigitaldivide.com/
[19]http://cbdd.wsu.edu/projects/global/lmi/index.html
[20]http://www.usaid.gov/our_work/economic_growth_and_trade/info_technology/special_initiatives/last_mile_initiative_overview.html
[21]http://horizons.wsu.edu/
[22]http://www.ed.gov/programs/comtechcenters/index.html

- The U.S. Department of Education's Technology Literacy Challenge Fund[23] gives grants to local school districts to upgrade their technology education. Although the funds can be used to purchase hardware and software, in many cases the emphasis has been on educating teachers how to use and teach about technology.
- The International Telecommunications Union and Cisco Systems have formed a partnership to fund and operate the Internet Training Centre Initiative for Developing Countries.[24] The organization provides training in network design, installation, and management. In addition, the organization focuses on educating women in developing countries.[25]
- The Open Society Institute operates its Information Program[26] throughout the world to enhance the access to knowledge, communication, and freedom of expression in poor and developing areas. Rather than supplying direct grants to establish technology facilities, it supports local policymakers and encourages them to use local resources to increase technology access.

Do They Work? The programs discussed in the preceding section are merely a sampling of the many ongoing projects attempting to bridge the digital divide.[27] Most focus on improving technology infrastructure and educating people in technology use, but the question remains: Do these efforts work? Are they having the desired impact? In most cases it's still too soon to tell. Over the next decade we'll be looking to see what happens, including trying to find answers to the following questions:

- Will people use the information they gain by having access to the Internet to start businesses?
- Will exposure to the Internet give people the motivation to further their educations? Will that education translate to better jobs and better economic situations?
- Is the world exposed by the Internet out of reach of many people in developing countries? Does knowledge of what they can't have cause frustration and make their living situations more intolerable?

---

[23] http://www.ed.gov/Technology/digdiv_projects.html
[24] http://www.un.org/Pubs/chronicle/2003/webArticles/012003_women_bridge_digitaldivide.html
[25] Although there is virtually no difference between technology use by men and women in developed countries, the same is not true in developing countries, where existing users are overwhelmingly male.
[26] http://www.soros.org/initiatives/information/about
[27] To find more projects that are attempting to bring information technology to those who don't have access, try a Google search on "bridge digital divide" or "bridging digital divide."

WHERE WE'VE BEEN

Access to the Internet isn't allocated evenly throughout the world. Western civilizations, including the United States, Canada, Western Europe, and Australia, have the highest penetration rates. Although high-population countries such as China and India have more users by raw count, those users are a smaller percentage of the population.

Barriers to Internet access can be economic (the person cannot afford Internet access), political (government policies or other activities restrict access), or personal choice (the person does not want to access the Internet). People without technology access or knowledge may find their job opportunities limited. They may also not have access to the latest news or the wide range of business opportunities that the Internet provides.

The gap between those who have access and those who don't is known as the "digital divide." Although there is some controversy as to whether the digital divide is growing and permanent, there are many programs throughout the world that are attempting to provide technology access. Most of these programs are too new to determine whether they are having the desired effect.

THINGS TO THINK ABOUT

1. As of this date, what is the Internet penetration in North America? In what country is Internet access growing the fastest? (To find the answers, you will need to do some Internet research.) If you were looking to start an ISP, in which country would you invest, based on potential growth in the number of users? Why?

2. Imagine that you are the CEO of a company that administers a major search engine. A foreign government has asked you to censor the results of searches that originate from within its borders using criteria that the government will supply. How do you respond to this request? Why?

3. Assume that you are confronted by a person who is afraid of computers and also concerned that he or she can't learn to use them. What can you say or do to help this person become comfortable and competent with using a computer to access the Internet and use e-mail?

4. You have been given $5 million for a project that will attempt to bridge the digital divide. Where will you concentrate your efforts? Describe what this project will do and how it will benefit those who participate.

5. Computing is primarily a visual medium. Typical computer use today also requires at least some physical agility (to use the mouse and keyboard). What type of barrier do these requirements create for those with limited sight or fine motor mobility? After doing some Internet research, describe some adaptive technologies that can lower that barrier.

WHERE TO GO FROM HERE

Internet Access in Cuba

BBC News. "Cuba law tightens Internet access."
http://news.bbc.co.uk/2/hi/americas/3425425.stm

CNN. "Cuba tightens its control over Internet."
http://www.cnn.com/2004/TECH/internet/01/21/cuba.internet.reut/

"Cuba tightens grip on net access."
http://www.wired.com/politics/law/news/2004/01/61866

Internet Access in China

"Controls on Internet access, China." *http://www.technologynewsdaily.com/node/554*

Einhorn, Bruce. "No new Internet cafes in China." *http://www.businessweek.com /globalbiz/blog/asiatech/archives/2007/03/no_new_internet.html*

Hermida, Alfred. "Behind China's Internet red firewall."
http://news.bbc.co.uk/1/hi/technology/2234154.stm

"Internet access in China." *http://www.marketresearch.com/map/prod/1524755.html*

"Internet access in China." *http://www.iht.com/articles/2006/02/20/opinion/edlet.php*

Reese, Brad. "Uncensored Internet access in China."
http://www.networkworld.com/community/node/12810

Wiley, Richard. "China overtaking US for fast Internet access as Africa gets left behind." *http://business.guardian.co.uk/story/0,,2102517,00.html*

Digital Divide

"Bridging the digital divide."
http://www.riverdeep.net/current/2002/01/011402t_divide.jhtml

Digital Divide.org. *http://www.digitaldivide.org/dd/index.html*

Digital Divide Network. *http://www.digitaldivide.net/*

Nielsen, Jakob. "Digital divide: the three stages."
http://www.useit.com/alertbox/digital-divide.html

OLPC Project

Glaskowsky, Peter. "OLPC battery life-an update."
http://news.com/8301-10784_3-9768920-7.html?part=rss&subj=news&tag=2547-1_3-0-5

Economics and Work **6**

WHERE WE'RE GOING

In this chapter we will

- Be introduced briefly to the history of U.S. economics.
- Discuss the places where manufacturing takes place today.
- Consider whether people living in a service economy are better off than they were in a manufacturing or agricultural economy.
- Talk about businesses spawned by technology and the characteristics of those that succeed and those that fail.
- Look at how our shopping behavior has changed because of technology.
- Discuss how some industries have adapted to changes in technology.
- Examine the impact economics have on innovation.

INTRODUCTION

For most of us who use the Internet, many of the impacts of the Internet on financial matters are obvious: We do our shopping and banking online. But technology has changed economies—and specific industries—in far more fundamental ways. In this chapter we will look at the impact of technology on economics, with a focus on the United States. We will put these effects in their historical context, looking at how agriculture, manufacturing, banking, and service industries interrelate and how technology has changed those relationships. We will also consider how technology has changed how and where we work.

NORTH AMERICAN ECONOMIC HISTORY

The North American economy (the United States and Canada, for purposes of this discussion) is arguably the largest economy in the world. It has changed significantly from its inception. Although geography has played a part in differentiating the U.S. and Canadian economies, for the most part they have moved together, sharing booms and busts. To help put the impact of technology on that economy into perspective, we will begin with a short history.

Agriculture

Initially, the British colonies in North America were financed by the British charter system, through which the British government granted charters to those who could invest in the new lands. A charter gave the owner land and some economic, judicial, and political rights over the land. However, most colonial charters did not pay off rapidly and frustrated owners released their charters to the settlers.

The departure of charter holders allowed settlers to create their own economy. Many Canadian colonists made their living through trapping, hunting, fishing, and logging, many selling their furs to the Hudson's Bay Company.[1] Those in the southern part of the country were also farmers.[2] Some colonists in what is now the United States also participated in trapping, hunting, and fishing, but most lived on small farms. There were a few large plantations in the South, but by in large the agriculture provided only a subsistence living. Luxuries had to be imported, generally from England.[3]

The settlement of the French Canadian colonies was heavily influenced by the Compagnie des Cent-Associés, which obtained its charter from Cardinal Richelieu. The charter gave the company's founders not only economic control of the lucrative fur trade, but also control over the expansion of France's colonies in North America. The Compagnie, however, suffered from the predations of privateers on their ships and therefore did not live up to its responsibilities of bringing a large number of settlers to the colonies. The French Canadian economy became independent of the Compagnie in the mid-1600s when the company was dissolved.

The type of commercial activity found in a specific location in the North American colonies was to a large extent determined by the geography and climate of the area. Canada had less land suited for farming than the United States. In particular, because of its northern location, it did not have the climate necessary to grow cotton or tobacco, two of the primary crops of the southern United States. However, Canada had more area that supported large populations of fur-bearing animals and thus had a larger trapping economy than the United States. Canada also had larger stands of timber that were suited for commercial logging in the eastern part of the Canadian colonies.

---

[1] The Hudson's Bay Company was North America's first corporation. The company still exists, although it sells general merchandise rather than furs. To find out what the company is doing today, see http://www.hbc.com/hbc/.

[2] When we speak of farming, we mean the raising of both plants and animals.

[3] As you might remember, all those imports gave the British government a myriad of opportunities to collect taxes. The colonies, however, had no say in what taxes were imposed, which eventually led to a cargo of tea being dumped in Boston harbor.

The agricultural economy required some supporting industries, such as mills for grinding grain, sawmills, shipyards, and ironworks. As the colonies moved through their wars for independence and into the 19th century, the North American economy became even more localized. Large plantations dominated the economy in the south, growing food products, cotton, and tobacco. The north was dominated by small farms and cottage industries. Manufacturing of goods such as weapons or wagons was performed by small firms. Most people, however, made their living through agriculture, furs, or timber.

Industry

The Industrial Revolution began in Europe during the last decades of the 18th century. It was triggered by technological development, specifically the technology to mass produce many items that formerly had been made by hand. Manufacturing spread to North America, and by the middle of the 19th century the continent was generating around one-third of its income from that type of production.

Most of the early industrialization was in the northern portion of the United States (in particular, the northeast) and in southern Canada; before the Civil War, the United States retained its largely agricultural economy. Northern industry focused on necessities that had previously been produced by individuals: weaving of cotton cloth, shoes, knitting of wool clothing, and machinery. In Canada the digging of canals facilitated the increased movement of timber, furs, and farm products to manufacturers.

Industrialization was supported by the influx of European immigrants during the 18th and 19th centuries. Most of these immigrants came to North America with little more than the clothes on their backs and were usually willing to work in factories at relatively low wages. They really didn't have much of a choice: Discrimination against immigrants ensured that many of those who were educated were not hired in the high-level positions for which they were qualified.

The Civil War changed the southern U.S. economy significantly. The plantations were profitable because their labor costs were very low. With the absolution of slavery, many of the plantations cost too much to operate at a profit. The south had no choice but to follow the north and look at industry to generate income.

After the Civil War, industry continued to expand. The entire continent entered a period of rapid technological innovation. Many of the items we consider commonplace today—for example, cars, telephones, bicycles, elevators, and traffic lights[4]—were developed during the latter part of the 19th century. This is also the period during which oil and coal became the fuels of choice; they were relatively cheap to extract and the supply seemed endless.

Along with the change to factory production, the first half of the 20th century saw the rise of trade unions. Manufacturers that had relied on cheap labor were forced to take the needs of workers into consideration. Wages—and prices—rose. As immigrants from the previous century were integrated into North American society and were hired for more skilled positions, new immigrants came to take

---

[4]The typewriter and the phonograph were also invented during this period, but they have been replaced with even more modern technologies.

their places in the factories. Many of the new wave of immigrants came from Eastern Europe (for example, Russia and Poland).

North America remained largely a manufacturing economy through the 1950s. Even agriculture was affected by the change: large farms appeared, operated by corporations rather than families. The corporate farms could produce more per acre than small farms and became more profitable.

Fueled by unions, the wages of North American workers continued to rise, causing increases in the cost of manufactured goods. It was cheaper to purchase imports of items such as clothing and shoes; many electronic devices, such as radios and televisions, were also imported. The United States began to accumulate a trade deficit, including one with Canada, its largest trading partner.

The 1960s and 1970s were marked by significant price inflation across the continent. Military spending increased to support the Vietnam War. An OPEC (Organization of Petroleum Exporting Countries) embargo of the United States in the early 1970s led to significant rises in fuel costs and concomitant rises in the costs of manufacturing and transporting goods. In an attempt to stop inflation, the U.S. Federal Reserve Board increased interest rates.[5] The action had the desired effect, but it also slowed economic development; the country entered a major recession in the early 1980s. At this point, the Canadian economy was following the trends of the U.S. economy and therefore suffered from the recession as well.

Service

As the North American economy emerged from the recession of the 1980s, people were purchasing high-quality imported goods (for example, cars) instead of locally made items. The United States ran up a huge trade deficit, especially with countries such as Japan, and had a large national debt. The situation was made worse by President Reagan's "trickle down economic policies," which essentially increased spending and cut taxes.[6]

At the same time, a large portion of the manufacturing was moving out of North America, to Asia (especially China) and Mexico. The United States no longer manufactured televisions; it never manufactured cell phones. Although computers were assembled in U.S. plants, the parts that went into the computers were made overseas.

U.S. car manufacturers continued to exist, but they no longer dominated the U.S. car market. In fact, a number of foreign automobile companies opened assembly plants on North American soil (for example, Toyota and Honda). The cars were therefore assembled using North American workers, but the profits still went back to the foreign owners.

---

[5]The Federal Reserve controls the rate at which banks borrow money. Rates that banks then charge businesses and individuals for borrowing are based on this "prime" rate plus a few percentage points. When the prime rate rises, it costs more to borrow money and people earn more by saving. This means less money circulating in the economy and less purchasing. As demand goes down, so does inflation. At least, that's the theory.
[6]Trickle down economics benefited the wealthiest Americans. The idea was that the wealthy would invest more in their businesses, thus creating jobs and increasing wages so that the benefits would trickle down to the rest of society. The problem was that it just didn't work.

Family farm owners struggled even more. There was (and still is) competitive pressure from large corporate farms. In addition, farmland has become extremely valuable for housing, and farmers who cannot make a living from farming may find it financially lucrative to sell the land to a developer.[7]

At the same time, new businesses were emerging, businesses that didn't manufacture a tangible good or create foodstuffs but that provided a service, such as banking, retail, consulting (especially for management and computer systems development), and software development.[8] The number of people employed in and the revenue generated by such businesses increased significantly.

During the presidency of Bill Clinton, the U.S. economy produced a budget surplus. This did not, however, eliminate the trade deficit or the national debt. Today, the U.S. economy still suffers from a massive trade deficit (especially with China) and a large national debt. The Federal Reserve keeps inflation in check, although housing costs in some parts of the country are still disproportionately high. As the U.S. economy suffered from the trade deficit and rising oil prices and appeared to be heading toward a recession beginning in 2007, the Canadian economy seemed at that time to be withstanding the pressures and avoiding the downturn.[9]

WHERE DID ALL THE MANUFACTURING GO?

Given that the North American economy is dominated by service industries, where are the goods we need manufactured? Some are still made in North America (although such goods tend to cost more than comparable items from overseas[10]); many goods, however, are manufactured in Brazil, Mexico, and China and other Asian countries.

There are several ways to look at the distribution of economies around the world. One is by considering the distribution of the labor force: the percentage of the population involved in agriculture, industry, and service jobs. Data for a selection of countries can be found in **Figure 6.1**. These data aren't as straightforward as they might seem, however. It's true that the less developed a country, the greater the proportion of its labor force engaged in agriculture. Countries such as Burundi and Ethiopia are overwhelming agricultural.

However, India, which is the source of a significant amount of today's computer software development, has a rapidly emerging middle class and supports many service jobs, but the country is still largely agricultural. The data

---

[7]Why is it so difficult to break even as a farmer? Consider the cost of equipment (tens of thousands of dollars per piece), animals, animal feed, labor, and so on. Many small farmers borrow money from a bank each year before planting. They purchase seed and any necessary equipment and then hope that the crops bring in enough to pay back the loan and support the family until the next planting season, when the cycle begins again. Many small farmers consider it a good year if they break even.

[8]Although the software may be written in North America, or at least under the auspices of a North American firm, the production of disks and manuals to be sold may occur in Asia or Mexico.

[9]One indication of the changing relationship between the U.S. and Canadian economies in the mid-2000s was the relationship between the U.S. and Canadian dollars. Traditionally, the U.S. dollar had been worth more (about $1.30 Canadian), making many goods more expensive in Canada than in the United States. However, by 2007 the two currencies were at parity and Canadians were beginning to complain about being charged higher prices.

[10]One exception to this is automobiles. The majority of the world's most costly cars—Porche, Mercedes-Benz, Rolls Royce, for example—are foreign made.

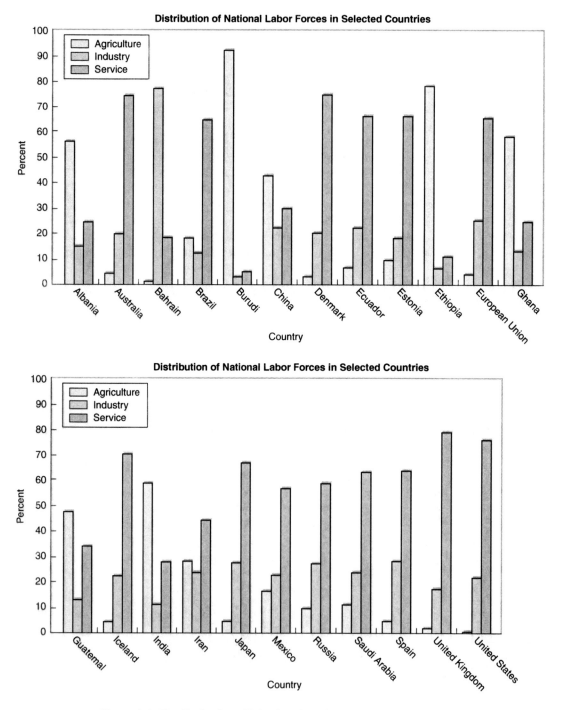

Figure 6.1 The distribution of labor in selected countries.

don't indicate the gulf between the educated middle class and the poor, some of whom live in rural villages and others who live on the streets in major cities. The poor often don't have a chance to get the education and opportunities they need to change their economic status.[11]

According to manufacturing statistics, the income generated by manufacturing in the United States has remained at about the same percentage of the gross national product (16% to 19%) and the number of workers involved in manufacturing has remained about the same. However, these statistics don't show the entire picture: The population keeps going up. This means that the percentage of U.S. workers with manufacturing jobs has actually declined (from 32% in 1947 to 11.5% in 2002).[12] Part of the decline in the percentage of manufacturing jobs is due to automation; another part is due to outsourcing, which is discussed later in this chapter. Ironically, automation has made each manufacturing worker more productive, keeping output from existing plants high.

China has become a significant source of manufactured goods. It has a large workforce with wages far below those in the United States, Canada, or Europe. Although the chart in Figure 6.1 indicates that the largest proportion of China's population engages in agriculture, the Chinese population is so large that the number of people engaged in manufacturing is considerable.

The quality of Chinese goods is generally high and consistent.[13] By 2006 China's industrial production was growing at a rate of nearly 23% a year.[14] In 2001 China exported only half the manufactured goods as the United States, However, by 2006 China became the world's largest exporter: The value of U.S. exported goods was $367 billion, whereas China's exports totaled $404 billion.[15] China has run up a huge trade surplus, whereas the United States has a trade deficit.[16]

AND ARE WE BETTER OFF?

In North America the service economy is a fact. This has brought both benefits and problems to the continent. First, we rely on other countries for some goods and if trade stops, we would encounter noticeable shortages of some classes of imported products. We wouldn't be able to purchase new televisions or cell phones; computer and automobile assembly would slow down because of a shortage of parts. Yet we are not alone in our dependence on other countries; trade

[11]The situation in India is even more complex than it first appears. Although the caste system has been legally abolished, a person's social status to a large extent determines opportunities for economic advancement. This makes it even more difficult, especially in large cities, to change the lives of those living in abject poverty.

[12]For more information, see http://www.jimpinto.com/writings/chinachallenge.html

[13]At the time this book was written, two major problems had arisen with Chinese goods. One was the use of lead paint on children's toys, causing a massive toy recall by companies such as Mattel and its subsidiary Fisher-Price. The second involved food. Chinese food preparation and packaging plants were not being held to stringent health standards, causing great concern for the athletes scheduled to attend the 2008 Summer Olympic Games in Beijing.

[14]From the CIA World Factbook, https://www.cia.gov/library/publications/the-world-factbook/rankorder/2089rank.html

[15]Statistics from http://www.manufacturingnews.com/news/06/0905/art1.html

[16]Although a trade surplus sounds like a good thing for a country, China may have some problems with its emphasis on international trade. It has ramped up exports at the expense of internal trade. Exports may not be able to grow much more, but to sustain its growth the Chinese economy needs to sell more. This may require taking a break from a focus on exports and spending time developing internal markets.

has become increasingly global, and there are very few nations that could exist without it today. The economies of the world are far more intertwined today than they have ever been.

Job Shifts

The service economy has also changed the type of jobs available to workers. Consider what has happened in automobile manufacturing. Initially, cars were assembled by hand. This required a large workforce, and once a worker became an employee of an automobile company, he or she was fairly certain to be employed for life. Then along came technology and automation. Fewer workers were needed on the assembly lines and layoffs began.[17]

Automation, however, also created new jobs: Someone had to operate and maintain the robots. The created jobs required more education than the more manual assembly line jobs, so the workers who were laid off were usually not qualified for the new positions. Instead, the jobs usually went to younger workers who had been trained in the new technologies.

The jobs created by a technology-driven service economy are either highly skilled, requiring education beyond high school, or they are unskilled. You can find a selection of jobs in each category in **Table 6.1**. Some of the jobs aren't totally new but have new requirements that are dictated by the use of technology.

As an example, consider how the job of a clerical worker has changed since the introduction of the PC. The people for whom a secretary works are likely to have PCs on their desks now. Managers send their own e-mail, eliminating the need for a clerical worker to type letters or take dictation. On the other hand, clerical workers need to be skilled in using PC applications and may be expected to perform desktop publishing to produce professional looking documents; he or she may also need to do Internet research to answers questions for an employer. It's no longer enough for a clerical worker to be able to type and take notes in shorthand.

Table 6.1 Selected Skilled and Unskilled Jobs in a Service Economy

Highly Skilled Jobs	Unskilled Jobs
Computer programmer	Fast-food restaurant worker
Electrical engineer	Retail salesperson
Systems analyst	
Clerical worker	
Technical support person	
Computer repair technician	
Satellite television installation and repair person	

---

[17]To be fair, part of the downsizing of the automobile industry is due to foreign competition. The percentage of U.S. purchases of American-made vehicles has declined since the days of the Model T Ford.

Unskilled jobs usually require some on-the-job training, although even those positions can't avoid using technology. Most fast-food restaurants, for example, are part of chains that keep track of sales and inventory using cash registers that are essentially specialized computers. The worker is presented with a touch-sensitive interface that includes a button for each menu item. All the worker needs to do is press the appropriate buttons to record the order. The computer terminal totals the order, the worker enters the amount tendered, and the terminal computes the change. Meanwhile, the order is sent to the server in the back of the restaurant, which displays the order on a monitor for workers to use when filling the order. Because training can be given on-the-job, a worker for this type of job needs to be able to read and count money, but little else.

Highly skilled jobs tend to be relatively well paid, whereas people in unskilled jobs often earn only the minimum wage. One concern of social scientists is the gap between rich and poor that this engenders. Data suggest that this gap is growing.[18] The United States was predicated on the idea that any person, with the right education, could rise out of poverty to lead a comfortable (though not necessarily rich) life. The widening gap between the rich and poor suggests that doing so is becoming more difficult.

Where We Work

One of the predictions of the 1950s that was made frequently about the year 2000 was that people would no longer travel to work, that we'd all be working out of our homes. For some people, telecommuting is a reality, yet others have increasingly long commutes to reach their workplaces.

A telecommuter is someone who works one or more days a week from a location (usually home) other than his or her employer's main place of business. The worker is connected to the business office by a data communications link. Not all jobs are amenable to telecommuting. For example, it is unlikely that a clerical worker could do much of his or her job at home. A computer programmer, however, is more likely to be able to work as well at home as in the office because a large portion of programming is solitary work. Jobs such as medical transcription or desktop publishing are also relatively solitary activities that lend themselves to working from home.

Only about 5% of the U.S. workforce telecommutes full time, although nearly 23% telecommute at least one day a week.[19] There are a number of reasons why more people don't work at home:

- As just discussed, not all jobs are amenable to telecommuting.
- Some employers don't trust their employees to complete their work without direct supervision.

---

[18] See http://freedomkeys.com/gap.htm and http://www.post-gazette.com/pg/05133/504149.stm
[19] See http://www.allbusiness.com/human-resources/employee-development-employee-productivity/892479-1.html and http://www.prnewswire.com/cgi-bin/stories.pl?ACCT=104&STORY=/www/story/07-19-2006/0004399403&EDATE=

- Employees who aren't seen in the office have less chance of job advancement and promotion because they aren't in the office enough for supervisors to see the value of their work.
- Although a job could be performed independently, interaction between colleagues—formal or informal around the water cooler—provides an important source of encouragement and problem solving.
- The home environment isn't conducive to work. Anyone who thinks he or she can provide child care and work at the same time is deluded. Successful home workers either work when there are no children at home or hire a day care provider—either in the home or in a day care center—to watch the children during the workday.

At the same time, some workers—especially those who work in heavily developed urban areas—are moving further from their workplaces. Housing in many major North American cities has become prohibitively expensive; it may be cheaper to live in the suburbs and to take the train or drive to work, even with high gasoline prices. In 2003 the U.S. Census Bureau found that those who worked in New York City had an average commute of 38.3 minutes.[20] Averages, however, can be deceiving. There are people from the Mid-Hudson Valley who have a one-way commute to Manhattan of nearly two hours! In Canada, where the population density is less than in the United States, even more people have long commutes.[21]

However, if land is available for development, the situation may be somewhat different. When companies build major campuses in formerly rural areas, housing often grows up around them. For example, Intel is headquartered in Beaverton, Oregon. The surrounding area has become a sea of housing, from apartment complexes to single-family homes. Services, such as shopping, are slower to enter the area than housing, and early residents have to travel further for the goods they need than they would if they lived in a more established area. Nonetheless, the conversion of farmland to business and residential use does allow many people to live affordably and close to where they work.

Outsourcing One of the biggest problems with the North American service economy is that wages in North America are quite high compared with much of the world. Many service jobs—such as those involving consulting or software development—require significant education and thus command high salaries. In an attempt to cut those costs, a number of businesses have outsourced jobs, sending them overseas to countries such as India and Mexico where labor costs are lower. Software development and computer technical support have lent themselves to significant amounts of outsourcing.[22]

---

[20]These data come from http://www.census.gov/acs/www/Products/Ranking/2003/R04T160.htm
[21]The United States has approximately 10 people per square mile, whereas Canada has only eight. For more information about Canada and telecommuting, see http://www.ivc.ca/canadianscene.html
[22]Outsourcing technical support backfired on Dell Computer. Users were so unhappy that the company had to bring the technical support back to the United States.

As outsourcing became a standard practice in the first years of the 21st century, many people were afraid that the result would be significant unemployment in the North American technology sector. Because companies would be able to hire people with the same expertise for less money overseas, North American workers would need to accept lower salaries if they wanted to work at all. Some jobs have indeed been lost, but the wage losses and wholesale unemployment of North American technology professionals has not occurred. There are a number of reasons why:

- Most companies that outsource want to keep the management of outsourced projects in-house. This has created additional jobs for project managers.
- Managing a project where the managers are in one time zone and the workers are in another, 12 hours away, is difficult. In some cases, outsourcing just hasn't been workable and was discontinued.
- Customers may be unhappy with contacts with outsourced workers. To avoid losing business, the company may have to bring the jobs back to the United States, where customers can interact with native English (or French) speakers.
- The threat of outsourcing has kept some technology professionals from seeking available jobs for fear that there is little job security because their jobs could be outsourced at any time. Companies then find it difficult to recruit North American staff and must continue to pay high salaries to those they are able to hire.[23]
- Some North American workers don't want to perform routine tasks such as transaction processing and technical support. They are perfectly happy to see those functions go overseas so they can engage in more interesting projects.
- Some of the outsourced jobs are new jobs that aren't replacing existing American positions.
- Enough jobs have been created within North America to compensate for many of the outsourced jobs. These tend to be the more complex, higher paid jobs that aren't amenable to outsourcing.

The preceding does not mean that outsourcing is not increasing—it is—but it does mean that outsourcing hasn't had the devastating effect that many assumed it would. How fast is it growing? Consider the chart in **Figure 6.2**, which illustrates the plans of organizations that outsource.[24] The total height of each bar represents the percent of organizations that outsource a particular IT activity. Notice that nearly 60% of companies outsource software development and that 26% plan to increase the amount of outsourcing they do. In fact, companies plan to increase the amount of their work that is outsourced across all types of IT activities.

---

[23] http://www.cfo.com/article.cfm/9569700?f=search
[24] These data come from http://www.computereconomics.com/article.cfm?id=1161

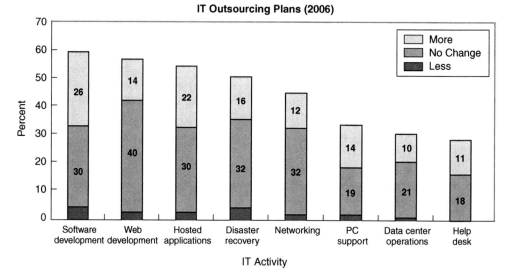

Figure 6.2 Outsourcing plans of companies that currently outsource.

HOW WE WORK: THE PAPERLESS OFFICE

In 1975 an article in *Business Week* predicted that we would be doing most of our business electronically by 1990.[25] This idea is known as the "paperless office." Ideally, IT would make it possible for all data to be stored and retrieved electronically; all correspondence would be electronic. In fact, Alvin Toffler, a well-known futurist, wrote in 1984 that "making paper copies of anything was a primitive use of electronic word-processing machines and violated their very spirit."[26]

The idea seems to have a great deal of merit. A business could save storage space by eliminating paper files; it could save money on storage space, storage equipment, and paper purchases. The idea of saving trees and saving the energy needed to produce paper is very attractive, especially when we consider the current push to reduce an organization's "carbon footprint." However, although technology makes it possible to use less paper, most businesses are far from a purely paperless office.

Through at least the early 2000s, paper use actually increased.[27] The introduction of high-speed photocopiers and laser printers made it easier to produce more paper documents more easily. E-mail and the Web didn't cut down on paper either; people often would print out their e-mail and the Web sites they visited.[28] But why do people print out documents that they receive online? That seems to be significantly counterproductive.

[25]"The Office of the Future." *Business Week*, June 30, 1975.
[26]Toffler, Alvin. *The Third Wave*. New York: Bantam, 1984.
[27]See http://earthtrends.wri.org/features/view_feature.php?theme=6&fid=19
[28]Have you ever been instructed by a Web site to print out a confirmation page after you've made a purchase? The Web itself is contributing to excess paper use!

The problem is that the theory of the paperless office doesn't take into account the way people feel about paper. Some people don't think that a document is "real" unless they can feel it in their hands; paper is often easier to read than a computer screen; paper is portable, can be folded to fit into a pocket, and can be marked up with a pen. There is therefore a great deal of resistance to doing away with paper documents entirely.

Other factors also make is difficult for a business to go paperless. Businesses must convert their paper-based records to digital form. Not only is this a time-consuming and often expensive process, but the optical character recognition (OCR) software that is used to translate a scanned image into editable text is prone to errors; converted documents must be compared to the printed original and corrected by hand. In addition, business partners of an organization attempting to go paperless may require paper documents.

Paper use is now going down, in part the result of people not printing as many e-mail messages, and some businesses have been able to convert to paperless operations.[29] However, the idea that a business can be totally paperless largely has been replaced by the goal of simply using less paper.[30]

BUSINESSES THAT MAKE TECHNOLOGY: RISE AND FALL OF THE DOT-COMS

One of the most interesting episodes of the economics of technology involves firms that actually create technology. Known as the "dot-coms," such businesses are designed to create new hardware and software products and to find new ways to use the Internet to make money. For a time, it seemed that such firms were golden.

In the 1900s, venture capitalists believed that anyone who had an idea for a technology- or Internet-based business was going to succeed and make lots and lots of money. The venture capitalists therefore poured millions of dollars into startup firms, many of which had barely the flicker of an idea of what they were going to do. Employees received unrealistically high salaries and amazing perqs. For example, some companies brought in catered meals for the employees, gave employees expensive cars, and paid for health club memberships. However, some dot-coms had begun without well-conceived business plans; they were working with an idea but had little knowledge of how to generate income from that idea.

By 2000 many of the dot-coms were in trouble. They had spent their first round of venture capital investment and had no product or service to show for it. Because so many of the dot-coms still had no income, venture capitalists were reluctant to sink more money into them. Funding dried up and the dot-coms that were operating in the red had to close. This led to a period of job shortage in the IT industry, as former employees of failed dot-coms flooded the market, all looking for jobs at the same time.

---

[29]See http://itmanagement.earthweb.com/article.php/3703551 for examples.
[30]See http://www.techdirt.com/articles/20060109/0040253.shtml for an online discussion of this trend.

You can find a summary of some of the top dot-com failures in **Table 6.2**.[31] Some were good ideas, such as home grocery delivery, but were poorly implemented or before their time.[32] Others were just bad ideas, such as the Web currency idea spawned by Flooz.com. Theoretically, a customer would send Flooz.com money, which could then be used like a gift card with participating online merchants. But Flooz.com was a startup and people were unwilling to trust it with their money.

The refusal of venture capital to put more money into failing dot-coms sent ripples throughout the world's economy. As the money dried up, the inflated prices of dot-com stock went down significantly. On March 20, 2000 the NASDAQ stock market, where many technology company stocks are traded, closed at just over 5,000 points, a new high. But just five days later, it had fallen to just under 4,600. It continued to fall through part of 2003, reaching a low of slightly under 2,000 points.

The failure of so many dot-coms does not mean that a new business cannot grow out of the Internet. In fact, a few have done so and have been spectacularly successful (see the section, Businesses Spawned by the Internet, for details).

Table 6.2   Dot-Com Failures

Name	Years of Operation	Type of Business	Amount of Money Lost (in U.S. Dollars)
Webvan	1999–2001	Home grocery delivery	1.2 billion
Pets.com	1998–2000	Pet supplies	82.5 million
Kozmo.com	1998–2001	Fast home delivery service for items such as snacks and DVDs	280 million
Flooz.com	1998–2001	Online currency as an alternative to credit cards	35 million
eToys.com*	1997–2001	Online toy store	166 million
Boo.com	1998–2000	Online fashion store	160 million
MPV.com	1999–2000	Sporting goods endorsed by sporting celebrities	85 million
Go.com	1998–2001	A portal to Disney company sites	790 million
Kibu.com	1999–2000	Social networking site for teenage girls	22 million
GovWorks.com	1999–2000	Portal for working with municipal governments	15 million

*After its bankruptcy in 2001, eToys was purchased by KayBee toys. It is now independent again and can be found at http://www.etoys.com

---

[31] See http://www.cnet.com/4520-11136_1-6278387-1.html
[32] Home grocery delivery seems to work best if a brick and mortar store offers the service. The store's inventory is online and you place your order online. Then a store employee shops for you, using your grocery list. Finally, a store employee makes the delivery.

The Ripple Effect

The effect of the rise and fall of the dot-coms has not been limited to those who were working in the industry when the major failures occurred. In large part because of the media coverage of the dot-com surge and eventual meltdown, the number of undergraduate computing majors has fluctuated wildly. As you can see from **Table 6.3**, the percentage of entering freshmen selecting computer science as a major rose at a relatively steady rate until 2000 (when the failure of the dot-coms was making the evening news), after which it dropped off precipitously. Although there has been a very slight increase in recent years, it seems unlikely that enrollments will again reach the levels seen in 2000.

This ia a problem because the number of students earning degrees in computing is directly related to the number of people able and available to take technology job. According to the U.S. Bureau of Labor Statistics, the number of jobs in the design and implementation of computer systems will grow nearly 40% by 2016; the number of system maintenance, security, and other related jobs will grow nearly 80%.[33] At the same time, the number of students graduating with computing degrees will not keep pace with the need. The memory of the dot-com bust is still fresh in the minds of parents of college students: In some cases, they are actively steering their children away from computing careers because they believe there will be no jobs for them when they graduate. Therefore it seems likely that we will be facing a shortage of IT workers by 2010.

Table 6.3 The Percentage of Incoming College Freshmen Selecting Computer Science Majors

Year	Men	Women	Total
1994	3.0	0.9	2.0
1995	3.5	1.0	2.4
1996	4.5	1.1	2.6
1997	5.0	1.4	3.0
1998	5.6	1.5	3.4
1999	6.5	1.4	3.7
2000	6.4	1.4	3.7
2001	6.0	1.2	3.4
2002	4.5	0.6	2.5
2003	3.5	0.4	1.6
2004	2.9	0.4	1.4
2005	n/a	n/a	1.6

From http://www.cra.org/CRN/articles/may05/vegso

---

[33] See http://www.bls.gov/oco/oc02003.htm

If the predicted shortage does materialize and is reported widely in the popular press, we may see another upswing in computing major enrollments. It takes four years, however, for any increase in entering freshmen to affect the marketplace. While the business world waits for a new crop of workers, salaries for existing technology professionals are likely to rise. The rise will encourage yet more students to choose technology majors and a glut of workers will flood the market, driving down salaries. The point is that because popular perceptions of the health of the IT marketplace have such a large effect on the number of people entering the field, the supply of IT workers is likely to remain cyclical, at least for the foreseeable future.

BUSINESSES THAT MAKE TECHNOLOGY: THE ECONOMICS OF THE SOFTWARE INDUSTRY

The software industry has been in the news a great deal in the past 10 years, especially the antitrust suit brought by various governments against Microsoft. The difficulty facing courts and consumers is the unusual nature of the economics of software development and sale. The software market is, by its very nature, not a pure competitive market.

Consider first what it costs to sell a software package. Every software package has development costs, even if they are just the time put in by an individual enhancing an open source program. Once the development costs are spent, it costs almost nothing to create a unit of the software to sell. An economist would say that the "marginal cost" of software—what it costs to produce one additional unit for sale—is effectively zero, regardless of the quantity produced.[34]

In a pure competitive market, the selling cost of a product usually is the same as the marginal cost. But if that were true of the software market, all software would be free. Pricing strategies are therefore different for software than for products with a non-zero marginal cost. Software is priced by assigning a cost to the value the consumer places on the software. Another way to look at this is to say that software developers charge "what the traffic will bear."

As mentioned at the beginning of this section, the software market is not a simple competitive market, in which a large number of firms make the same product and compete for the same pool of customers. In contrast, when two software developers create word processors, the products are different and consumers usually choose only one. Each developer has a monopoly over the sales of its own products. This type of market is known as a "competitive monopoly."

Competitive monopolies are not necessarily a bad thing, and there are many of them in our economies. Books and music are two good examples: Each publisher has monopolistic control over the sales of its product, the products are differentiated but serve an equivalent function, and the consumer is likely to choose one product of a specific type over another. For example, if you have the money to

---

[34]Yes, it does cost money to duplicate media and prepare a physical package for sale, but economists still consider the marginal cost of software to be zero. Indeed, it is truly zero when you download software from the vendor's Web site and receive an electronic manual.

buy one hardcover book, you will probably consider several books from different publishers and then settle on the single title you want to purchase.

However, competitive monopolies are susceptible to price fixing, where the few companies in the market get together to set prices for their products. Price fixing is illegal in the United States and Canada because it is anticompetitive and thus works against free trade. Price fixing is unlikely in the software market because of the nature of the companies involved. When you are choosing an operating system for a desktop computer, for example, you can have Windows or some flavor of UNIX.[35] Users are most likely to choose only one operating system platform to the exclusion of all others.[36] Therefore the companies seeking market share want to keep prices as low as they can and still make money.

A practice that is legally murkier than price fixing is "tying," through which a company requires consumers to accept one product to be able to purchase another. In fact, the U.S. government based its 1998 antitrust suit against Microsoft on the tying of a Web browser (Microsoft's Internet Explorer) to an operating system. The basic argument was that by tying the browser to an operating system that most users would purchase, Microsoft was stifling competition in the browser market. The result was not what the U.S. government wanted: The initial verdict that required Microsoft to be split into two companies—one for applications and one for the operating system—was overturned. Instead, Microsoft was required to unbundle the browser from the Windows operating system. For the most part, Microsoft was left to continue business as usual in the United States. However, Microsoft is still dealing with lawsuits brought by other countries and the European Union.

The software market does show some characteristics of traditional competitive markets. There can be significant brand loyalty: Some consumers choose software from a specific vendor because they identify with the company and have used its products in the past. In addition, once a consumer has invested in an operating system and application software that runs under that operating system, it is very costly to switch to a different platform. These two factors tend to decrease the amount of competition in the market.

Why don't we have more software developers? Adobe rules the graphics world just as Microsoft rules the operating system and office suite markets. Why doesn't Adobe have serious competition? There are two main reasons. The first is that at this point most people who buy graphics software have already done so; there are far fewer new purchases than upgrade purchases. This provides a significant "barrier to entry" to a new firm, which must not only fund the development of a new product but must convince existing users to switch. Unless the new product provides significantly enhanced features and ease of use, people are unlikely to do so.

---

[35] Mac OS X is a version of UNIX, just like Linux.
[36] The major exception to this statement is the use of Windows on Intel-powered Macintoshes. The primary operating system is Mac OS X, but the computer can operate with Windows or Windows can run as a virtual machine within Mac OS X.

Second, there has been a great deal of consolidation in the software market. Adobe, for example, bought competitor Macromedia and folded many of their products into the Adobe line. The trend has been toward fewer firms that develop software rather than more. When such consolidation occurs, the number of people employed in the industry goes down. Although not all those who work for the acquired firm are laid off, there is usually a significant amount of overlap in staffing, and those in duplicated positions often find themselves out of a job.

Finally, we should consider the effects of software piracy on the market. During the 1980s, when PCs became widely used, software was sold on copy-protected disks. At the time, there was a burgeoning underground traffic in programs that could break copy protection. Given that digital copies are identical to the originals and that software in the early days of desktop machines was quite expensive, there was a lot of incentive for people to crack copy protection and make pirate copies of the software that could be sold or given away. As far as software companies were concerned, each pirate copy represented a lost sale.

Ultimately, the concerns of business users over the inability to make backup disks of mission-critical applications eventually led to the removal of copy protection. Software developers also began to lower the prices of software, making piracy less profitable.[37] Nonetheless, there is still significant software piracy, and, despite the efforts of organizations such as the Business Software Alliance, much piracy goes undetected. Given that we can never know how many of the people with pirated copies would buy the software if a free copy weren't available, it is difficult to assess the true impact of software piracy on the price of software.

SUCCESSFUL BUSINESSES SPAWNED BY THE INTERNET

Many of the successful businesses that operate on the Web also have brick-and-mortar stores. Department stores (e.g., JC Penney, Sears, Target, Kmart, Nordstroms) have found that it's profitable to sell online, reaching a large consumer base without the expense of printing and mailing a large catalog. Traditional catalog retailers (for example, L.L. Bean and Lands End) have also found that it is cost effective to support shopping over the Internet. But there are some businesses that exist on the Internet without having physical world analogs. These are businesses that could not exist without the Internet. In this section we will look at what makes some of those businesses unique and why they have succeeded while so many dot-coms failed.[38]

---

[37] The idea of lowering prices to defeat piracy didn't start with the computer software industry: It began with feature films. Movie piracy was rampant in the years after the VCR debuted, when prerecorded movies cost upwards of $80. Then, Paramount released *The Wrath of Khan* at $19.95. Suddenly, a prerecorded tape didn't cost much more than a pirated tape and the quality was significantly better. The strategy worked: Sales volume soared and the pirates were cut out of the market. Digital piracy of movies and music is a somewhat different issue and is discussed in Chapter 11.

[38] This discussion purposely leaves out social networking sites such as MySpace and Facebook. They'll be covered in Chapter 7, where we look at their impact on social interactions.

eBay and PayPal

eBay is a difficult business to categorize: It has no product. eBay acts as a middleman for people selling goods, although unlike a traditional auction house, it never handles the items being sold. When you pay eBay listing fees and final value fees, you are paying for the right to use eBay's computers to advertise what you have for sale, to conduct the sale, and to communicate with the purchaser. All the interactions take place through eBay's infrastructure; sellers and buyers rarely meet and it is up to the seller to get items sold to the buyer. eBay simply couldn't exist without the Internet (or, at least, without some sort of world-spanning network).

eBay was founded in 1995 by Pierre Omidyar, a computer programmer. He called it AuctionWeb and launched it as part of his personal Web site. He was as surprised as could be when he executed the first sale of a broken laser printer. By the fall of 1996, eBay existed as a real company with an employee and a president. (The name change from AuctionWeb to eBay[39] occurred in 1997.) Since that time, eBay has expanded to allow users to run their own virtual stores off the eBay site; there are people who generate their entire income from eBay sales. eBay Express, which launched in 2006, supports more traditional sales.

In 2002 eBay purchased PayPal, another business that relies on the Internet for its very existence. PayPal acts as a financial middleman, providing fund transfers between buyers and sellers. PayPal allows online merchants to accept credit card payments without a direct merchant account; it also provides protection for purchasers in that the seller never sees the buyer's funding source. Sellers maintain a PayPal account into which all funds paid to them are deposited. They can then use a PayPal debit MasterCard to pay for items using those funds at any place that accepts MasterCard; they can also transfer the funds to a personal or business bank account. PayPal makes money by charging a seller a percentage of funds received.

Legally, PayPal isn't a bank. It neither makes loans nor pays interest on the funds in user accounts. It is a clearinghouse for funds transfers, much like eBay is a clearinghouse for auctions. It has no brick-and-mortar business presence. Without the Internet, PayPal would not exist.

Amazon.com

Amazon.com is one of the dot-coms that "made it." After years of red ink, it began to show a profit in 2003. Amazon.com began as an online bookstore in 1995, offering steep discounts to make its books cheaper than those of physical bookstores, even when shipping was included. Today, Amazon sells various types of entertainment media, electronics, kitchen items, tools, garden items, clothing, jewelry, and sporting goods. It also handles online sales for other companies, such as Toys-R-Us, CD-Now, and Borders Books.

Amazon maintains a network of resellers to handle products that it may not carry. For example, if a book is out of print, one of the bookstores associated with Amazon may have a copy. The resellers ship directly to purchasers and pay Amazon a percentage of the purchase price in exchange for having their inventories included in amazon's listings.

---

[39] Omidyar's consulting firm was named Echo Bay Technology Group, but when he went to register the domain name, echobay.com was already taken. "eBay" was the next best thing.

Other retailers do as well as Amazon online, but Amazon is different in that it is the largest retailer that doesn't operate brick-and-mortar stores. Amazon does, however, have warehouses (ten in North America, nine in Europe, and four in Asia). It stocks the most popular items and can therefore ship them quickly.

Several factors combined to make amazon.com a success. First, the company was well funded so that it could weather several years of losing money while it built its customer base. Second, it offered merchandise discounts that couldn't be matched anywhere else and provided excellent customer service.

Google

When you want to find something specific on the Web and you don't know which Web site might have the information, where do you go? In all likelihood, it's to Google.[40] Google is the world's most widely used Web search engine. It became that way by using a method for indexing Web site content that continues to return more relevant Web pages than any other search engine.

Google was started in 1996 by Larry Page and Sergey Brin as part of a PhD project at Stanford University. The project was designed to produce a better Web search engine. Their theory involved indexing relationships between Web sites, believing that the more links a page had to it from other pages relevant to the search, the better the page would satisfy the search.[41] After testing using storage on Stanford's Web site proved that their theory would work, they registered the google.com domain name in 1997 and incorporated in 1998.

Google offers its services for free to users. It makes its money by selling advertising that appears on the side of a search result page. Notice that these ads are text only: no flashy graphics or annoying pop-up windows. But because so many people use Google, the sponsored ads pay off for their companies.

As it has grown, Google has expanded into other product offerings. It offers e-mail to anyone who wants it (g-mail[42]) and an extensive mapping service (Google Earth[43]) that can superimpose satellite photos on maps.[44] Google also provides its own business application suite (Google Docs[45]) that competes with Microsoft Office.[46] Another service, GrandCentral, provides a free telephone number in any North American area code the user chooses. When someone calls the number, GrandCentral forwards the call to a number of the user's choice.[47]

---

[40] Google is so much a part our world today that "google" (with the lowercase "g") has become a verb meaning "to search the Web for information." In other words, if you needed to research a project on the dot-coms, you would say "I'm going to google dot-coms."

[41] Other search engines at the time counted the number of times the search text appeared on a page; the more occurrences of the search text, the higher the page was placed in the search result.

[42] http://mail.google.com

[43] http://earth.google.com

[44] There is some concern that Google Earth could be used for surveillance of individuals, because in some cases the satellite imagery can show separate people. However, the information in Google Earth is available elsewhere on the Web (although not in the same accessible form) so perhaps the concern shouldn't be aimed at Google but at whomever put the surveillance satellites in orbit.

[45] http://docs.google.com

[46] For a relatively complete list of Google's products—and it's a rather extensive list—
see http://en.wikipedia.org/wiki/List_of_Google_products

[47] The beauty of GrandCentral is there is no charge for the forwarding of calls. Therefore if you have a phone number in an area code where it's a local call for the caller, there is no charge for the caller to reach you, regardless of the area code to which the call is forwarded. GrandCentral also forwards calls to multiple phone numbers simultaneously: No matter whether you are at home or at work or on the move, a forwarded call can reach you.

The basic level of service of each of these products is free, although in some cases users can upgrade to more extensive services for a small fee. For example, an individual can add global positioning system, or GPS, support to Google Earth for $20; organizations must pay more ($400). Probably the most amazing thing about Google and its free applications is that all of it is supported, and supported well, by advertising.

Why They Succeeded

Why are these companies successful while many of the dot-coms were not? What did they do that was different? Might those things act as a blueprint for startup businesses?

First, let's consider what these successful businesses have in common:

- An idea for a product or service that either isn't provided elsewhere or that the new business can provide better than most existing businesses
- A solid business plan
- Strong financial backing that is willing to wait for the business to build its customer base and become profitable
- Good management
- As the business grows, expansion into other, related products and services

The irony is that the preceding list can be applied to any new business, whether it is an Internet-only business, a brick-and-mortar–only business, or a combination of both. The bottom line is that successful Web businesses are different in the way they interact with their customers, but good business practices remain the same. Technology may have changed the way we do business, but it hasn't changed what makes a business successful.

HOW WE SHOP

It wasn't long after the inception of the World Wide Web that someone figured out a way to make money from it by selling stuff. However, electronic commerce (e-commerce, for short) actually began earlier. The first use of the term "electronic commerce," which was used in the 1970s, referred to using telecommunications to transmit business data, such as electronic data interchange (EDI) and electronic funds transfer (EFT). Electronic commerce expanded through the use of credit cards, for which transactions were authorized by a small terminal that placed a telephone call to the bank issuing the card.[48]

It's hard to be precise about when e-commerce spread into the consumer sector, providing a way for individual consumers to purchase goods. However, we do know that the Boston Computer Exchange began operations in 1982 to provide a venue for selling used computer equipment. Nonetheless, it was not until the mid-1990s that the Web became the vehicle for business-to-consumer shopping. The dot-com bust of the early 2000s weeded out many weak e-commerce businesses, and today the market (and its "blind hand") determines which e-commerce ventures succeed.

---

[48]The use of a modem and a standard telephone line to authorize credit card transactions is still state-of-the-art for many small businesses.

Like most things that involve technology, shopping online has both pluses and minuses. On the plus side,

- There is a wider selection of goods. It doesn't cost a great deal to have a Web-based store, so even small retailers can be accessible over the Internet.
- There is an easy comparison of features and prices among retailers.
- You can shop without leaving your home or office. For those who are constrained by physical problems or by time, shopping online may be the only possible way to shop.

On the minus side,

- The social aspects of shopping are lost. The importance of shopping as a social activity probably can't be underestimated. It is the only social interaction that some people have. For this reason alone, brick-and-mortar shopping isn't likely to disappear.
- It takes longer to ship an item from an online vendor than it does to get an in-stock item at a local store.[49]
- Goods cannot be examined before purchase. For example, although it's possible to order a new car online, most people would like to test drive several competing models before placing that order. In this situation, a consumer might visit dealers to do the test drive and then, once a decision has been made as to which car to purchase, shop online to find the best price.
- There is a risk involved with online shopping: Credit card numbers, personal information, and order details that sit on a merchant's hard disk are vulnerable to theft.[50] It is up to the retailer to provide security to prevent identity theft.

INDUSTRIES THAT HAVE EVOLVED WITH TECHNOLOGICAL CHANGE

Technology has spurred fundamental changes in some industries. These are traditional companies that have had to modify the way they do business so they can stay in business. We will take a look at banking, publishing, and agriculture in this chapter; we look at the music industry in Chapter 11.

Banking

Banking has a long history of being an early adopter of new technology. Anything that could help with computations could help banking: Even a mechanical adding machine was useful when there was nothing else available. As you can see in **Table 6.4,** banking began using computers for a major part of its operations as early as 1945.[51]

---

[49]The one notable exception to this statement is software. A number of software vendors make their products available for download. You pay for a serial number that unlocks the software. E-books, which we will discuss later in this book, are also usually available for immediate download.

[50]Contrary to public opinion, there is relatively little danger when a credit card number is traveling over the Internet. First, credit card numbers are encrypted. Second, there is so much traffic on the Internet that capturing specific packets that contain credit card numbers would be extremely time consuming and labor intensive. Third, all but the smallest messages are broken up into multiple packets, and not all of the packets may travel to their destination using the same route.

[51]There is a story floating around that Konrad Zuse's third machine, the Z3, survived the allied bombings of World War II and was used in a bank in Austria for a number of years.

To remain profitable, a bank must move money, whether it is making loans, clearing checks, or paying interest on financial instruments such as certificates of deposit. Before making a loan, a bank must be able to check the credit worthiness of the person or organization requesting the loan. The bank must then "service" the loan by sending monthly statements and crediting payments as they are made.

Credit Histories All our credit histories are stored on the computers of the three major credit-reporting bureaus; credit checks, that were once made by phone, are now made by computer. The companies that we pay transmit payment histories electronically to the credit bureaus; changes in our credit ratings are made faster than they were before the Internet provided easy access to the credit bureaus.

The idea of a computerized individual credit history that is readily accessible to those authorized to view it has therefore been a mixed blessing for most people. If our credit rating—that numeric score that the credit-reporting bureaus calculate based on how well we pay our bills—is good, we appreciate being able to get credit quickly. However, if our credit rating isn't great, we would rather that it wasn't recorded.

Check Handling The idea of check clearinghouses—places where checks from many banks are processed—dates back to the 18th century.[52] In the United States the primary check clearinghouse is the Federal Reserve. In Canada (and much of Europe), associations supported by groups of banks handle check clearing; there is no government agency involved in the process. Before the introduction of software that could read handwriting, an important part of the check-clearing process was manual. For example, the U.S. Federal Reserve hired people to sit at special terminals through which checks would pass. The operator would read the amount of the check and enter it into the machine, which would then print the amount on the bottom of the check in magnetic ink. Once so coded, the checks were processed automatically. Before 2004, although the funds associated with check processing were transferred electronically, the physical checks were distributed by truck to the banks upon which they were drawn. Checks were sorted at the branch (often by hand) and returned to customers in their monthly statements.

However, the Check 21 law[53] changed all that in the United States: Banks are now allowed to provide digital images of checks rather than returning the physical checks;[54] a law with similar provisions has been adopted in Canada.[55] Checks are photographed digitally and then made available to customers as small images on printed statements ("substitute checks") or as online images. According to Check 21, a substitute check is as valid a legal instrument as a canceled paper check. The result of this change is that checks clear much faster, in as little as 1 day.[56]

---

[52]See http://www.factmonster.com/ipka/A0001522.html for the possibly apocryphal story of how it began.
[53]See http://www.federalreserve.gov/paymentsystems/truncation/
[54]Most of the time, the original checks are shredded.
[55]See http://www.afponline.ca/pub/res/news/ns_20051202_image.html
[56]The banks certainly benefit from the faster clearing, and you do too, if you're depositing a check and your bank chooses to release the deposited funds to you as soon as the check clears. (Banks are not required to do so but may wait the legal time for holding deposited funds, which can be as long as five business days for out-of-state checks.) However, if you are trying to float a check before the next payday, the faster clearing could be a major problem!

Table 6.4 Uses of Technology in the Banking Industry During the Past Two Centuries

Years	Uses of Technology
Before 1846	Printed checks are introduced
1846–1945	*All the above plus:*
	Telegraphed stock prices (in particular, the use of a transatlantic telegraph cable to speed up international trades)
	Telephone communication between branches
1945–1968	*All the above plus:*
	Automated bank statements
	Check guarantee cards
	Automated checking clearing systems
	Magnetic ink character recognition (MICR) is introduced for coding the amount, account number, and bank identification on a check
1968–1980	*All the above plus:*
	ATMs
	Automated branch accounting
	General credit cards such as MasterCard and Visa
1980–1995	*All the above plus:*
	Telephone banking
	Electronic funds transfer
1996 and beyond	*All the above plus:*
	Online banking, including fund transfers, bill paying, and electronic statements
	Substitute checks (check images, authorized by Check 21 legislation in the United States)
	Debit cards

Several technologies have come together to make digital check systems a reality:

- Digital imaging: The checks are photographed using digital cameras so the images can be stored electronically.
- Optical character recognition: The amount of a check, the account from which it is drawn, and the bank on which it is drawn is read by computer.
- Massive amounts of external storage: External storage for computers continually increases in capacity and decreases in price. Because images consume more storage space than plain text or numbers, cheap high-capacity storage is essential for handling the substitute checks.

Online Banking The appearance of substitute checks has coincided with an expansion in online banking. Many of us no longer receive hard copy account statements: We do everything online, over the Internet. Statements, accurate to at least the prior business day, are available online; if we have multiple accounts, we can transfer funds between them online. We can pay bills online, even to those organizations that don't accept electronic fund transfers; the bank prepares the check and mails it for us. The beauty of this for most consumers is that it is all free.

The Internet has made online banking feasible because banks are freed of the expense and responsibility of maintaining the network itself. They can concentrate on developing their Web site functionality and keeping stored data secure.

Bottom Line Banks have historically been early adopters of new technology; the entire industry has upgraded with advances in computing technology. Any bank that doesn't keep pace simply won't remain in business. But although banking technology is typically very reliable, there is one major issue looming today: security.

Data are most vulnerable when they are stored in a fixed location, such as in a bank's storage array. As mentioned perviously, it is highly unlikely that packets traveling over the Internet will be intercepted in transit. But when data stop moving, they become a much easier target. Today, the fastest growing area of IT is security. Banks cannot be complacent about the technology they use; they must continue to find better ways to secure financial data against attackers. Not doing so will lose them customers (and, of course, their source of revenue).

Publishing

One of the first industries to undergo a major change because of technology was publishing: The laser printer and desktop publishing changed everything. To understand just what happened, we first need to look at the traditional way in which a printed book or magazine was produced:

1. The author used a typewriter to prepare the manuscript.
2. The manuscript was sent to the publisher, where multiple copies were made.
3. The manuscript was sent to reviewers, people who make comments on the content of the writing. The reviewers typed their comments and mailed them back to the publisher.

4. The author revised the manuscript. Rather than retyping the entire book, article, or paper, the author used a physical cut-and-paste process. He or she typed the revised sections, used scissors to cut the typed manuscript so that revised sections could be dropped out. Then, the revised copy was assembled, using as much new paper as necessary, using glue to paste the revised sections in the correct place.

5. The manuscript went to a copy editor, who made changes by hand using well-known copyediting marks.

6. The author reviewed the copyedited manuscript and returned it to the publisher.

7. The manuscript was retyped for the typesetting machine. The output was the text of the book, article, or paper in a long strip, the width of a single column of text. These were known as *galley proofs*, or simply *galleys*.

8. The author proofread the galleys and returned them to the publisher for correction.

9. Compositors cut the corrected galleys into chunks and pasted them to boards ruled with blue lines. Any illustrations were also pasted onto the "blue lines" at this point.

10. The blue lines were photographed to make the plates from which the item would be printed.

11. The plates were used to print the item. It was bound by machine.

This method was cumbersome and time consuming. For example, it was not unheard of for a scientific paper to take two years or more to see print.

That process has been changed by the use of technology:

1. The author uses a word processor to write the manuscript.

2. The author e-mails a PDF of the manuscript to the publisher.

3. The publisher e-mails the manuscript to reviewers, who e-mail their comments back to the publishers.

4. The publisher sends the reviews to the author, who then goes back to the original word processing files and makes any needed changes. Then, the author sends the files to the publisher once more.

5. The manuscript may be printed for copyediting. Although some publishers today use electronic copyediting, a significant portion do not. In that case, it's the only time the manuscript appears in hard copy until the final item is printed.

6. The copyedited manuscript is sent to the author for review. The author may send back physical pages with changes or may describe them in an e-mail.

7. The manuscript goes to the compositor, who makes the copy edits and uses desktop publishing software to create the layout of the text and illustrations.

8. Some publishers will send PDFs of the items pages to the author for review.

9. An image setter takes the final files—today, usually PDFs—and prints the plates from which the item will be printed.

10. The item is printed and bound.

The current procedure is not only faster than the original, but it uses fewer resources, especially paper. It saves mailing costs because, except for the copyediting phase for those publishers who still copyedit on hard copy, everything travels by e-mail. It is also less frustrating for the author, who can work with a word processor rather than doing revisions using the physical cut-and-paste method.

It took 10 to 15 years for most of the world's publishers to make the switch to electronic book production. However, even the most tradition-bound publishing houses realized that unless they modernized, they could not compete with those publishers who could produce items faster, cheaper, and—because once a correction is made in a word processed document it stays corrected—more accurately. Later editions of a book are also easier to prepare because the original doesn't need to be retyped as revisions are made: The author simply works from the original files.

One prediction that often was made in the 1950s about the year 2000 was that we would see the end of print materials: All books, magazines, and newspapers would be electronic. Although print publishers seemed eager to embrace electronic production of their products, they have been much slower to provide e-books. The reasons for this primarily are not technological. It's been more a question of demand: People just aren't ready to give up the physical feel of reading material and reading from a computer screen isn't as comfortable as curling up with a book. At this time, many scholarly publications and news services have migrated to the Web. The full text of a number of nonfiction books and books with expired copyrights are also available online, some of which are free and some of which need to be purchased.

In contrast, publishers have been eager to use online copies of selected pages from a book for marketing purposes. This lets potential purchasers read the table of contents and a few sample pages before they buy. If you look at many of the books in the inventory of major online retailers such as amazon.com and barnesandnoble.com, you'll see this marketing in action.

E-books Depending on what you read and to whom you listen, the publishing industry may be on the verge of another major change because of technology. The idea of an e-book, a book that is in electronic rather than print form and is read on a stand-alone reading device, has been around for a long time. It simply hasn't caught on: The readers have been too heavy and cumbersome to hold like a printed book and the amount of available content has been limited.

If amazon.com has its way, this will change. In 2007 it released Kindle, a paperback-sized e-book reader. At its release, amazon.com was providing 88,000 digital titles, costing no more than $9.99. Kindle can connect to amazon.com wirelessly through the cell phone network so users can browse, purchase, and download books without being physically connected to the Internet.

Will Kindle be the device that makes e-books a major factor in publishing? At this point, it's hard to know. There are some major limitations to the Kindle platform that make it less flexible than a printed book: Once purchased, books cannot be shared (unless you loan the entire Kindle device), given away, or resold. There is no way to print out pages from a Kindle book. These are all things that are easy to do with a printed book, and it remains to be seen if readers are prepared to give them up for the chance to store as many as 200 titles on a single, small device.[57]

Agriculture

Throughout history, the basic goal of human agriculture has been to get the most plant crop or animals out of a patch of land. In prehistoric eras farming was done with stone and stick implements with nothing more than human power behind them. Then along came the plow, first pulled by people and later by animals. Just as the arctic sled dog has been replaced by snowmobiles, the oxen and horses that pulled farm equipment have been replaced by machines. Computers and GPSs have also become a big part of today's agriculture. A recent tractor, sporting a GPS device in the middle of the top of its cab, is shown in **Figure 6.3**.

As mentioned earlier in this chapter, the trend in North American agriculture has been away from small family farms to large, corporate farms. One reason is the huge cost of the technology of agriculture; it's difficult for a family farmer to afford the technology, yet the small farmer can't compete without it. However, even with the number of farms decreasing, the productivity of North American farms goes up, mostly because the use of technology produces more crops and animals per acre. In 1950, one farmer could feed 15 people; by 1995, one farmer could supply enough food for 185 people (34 of whom live outside North America[58]).

A farm can use technology throughout its operations:

- Tractors are more comfortable for the driver, including touch-sensitive controls, such as those in **Figure 6.4**.
- Equipment that spreads seed, fertilizer, and pesticides can be programmed to vary the amount disbursed automatically, based on location.
- GPS can provide maps of farmland that can then be used as input to automatic seeding, spreading, and watering equipment.
- Computers track finances, provide online marketing opportunities, and manage the health records of livestock. The Internet has also created a means for the farming community to communicate easily.
- Radiofrequency identification (RFID) chips in animals' ears cannot only identify the individual animals, they also interface with automatic feeding systems. This is particularly useful for dairy farms, where each cow is fed in a specific stall while she is being milked. After identifying the cow as she comes into the barn, equipment can dispense the correct feed for each cow.
- Enhanced weather forecasting makes decisions about when to plant, when to fertilize, when to water, and when to harvest easier and more accurate.

---

[57]For an excellent analysis of Kindle and expectations for the device, see *Newsweek*, November 26, 2007, "The Future of Reading," page 57.
[58]http://www.dallasfed.org/research/pubs/agtech.html

Figure 6.3 A John Deere tractor equipped with a GPS device.

Figure 6.4 Touch-sensitive controls in a John Deere tractor.

The preceding list assumes that the farmer is upgrading traditional farming procedures. However, technology has gone farther: Bioengineered crops are being tested and grown. Some crops are bioengineered to be resistant to insect pests and disease. Others are more resistant to weather vagaries. Genetically engineered livestock are also available, such as cows that give more milk.

Bioengineering is not without controversy. One group of people object on religious grounds: They believe that tinkering with the genetic structure of living beings is interfering with God's plan for the Earth, that scientists are working in an area that ethically they should not. Another group warns that bioengineered foodstuffs (and the feed for food animals) haven't been tested for safe consumption by humans.[59]

Can a farmer make a living without using advanced technology? Yes and no. The Amish certainly do it. However, the Amish are mostly concerned with feeding themselves. Although they aren't averse to making a profit, that isn't their primary goal; they don't see their farms as a strictly business venture and they do not compete directly with corporate farms. They have no need to maximize the amount grown or animals raised per acre. On the other hand, a farmer who farms to make a living must compete with those who do use technology; the only way to remain competitive is to adopt the same technological aids that everyone else is using.

---

[59]For a discussion of the dangers of bovine growth hormone, which causes cows to give more milk, see http://www.shirleys-wellness-cafe.com/bgh.htm. An opposite position—that bovine growth hormone is harmless—can be found at http://www.fda.gov/bbs/topics/CONSUMER/CON00068.HTML

ECONOMICS AND INNOVATION

To this point, we've assumed that innovation just *happens*, but—as we discussed in Chapter 1—someone has to pay for it. The corporations that operate think tanks have to pay for facilities, labs, and employees. Whether research is theoretical or applied, the bills need to be paid by someone. Innovation therefore often relies on the financial stability of the organization funding the research effort.

Economics also affects innovation when new products conflict with existing products and, at least in the short run, mean a reduction in corporate revenue. It is also not uncommon for competing technologies to hit the market at about the same time: In some cases, only one will survive and it won't necessarily be the best technology. In this section we will look at both of those impacts of economics on technological innovation.

Can Economics Stifle Innovation?

Do economic concerns ever stifle innovation? When corporate profits are at stake, they can and do. Consider, for example, the search for an alternative to gasoline to power cars and trucks. Considerable research and development is occurring throughout the world to find a way to fuel our vehicles without putting emissions chemicals into the air. To understand where the research is happening, we have to look at who stands to win and who stands to lose when the world moves from fossil fuels to some other fuel source. The biggest losers will be the big oil companies, unless they can retool themselves to sell the new fuels. Oil companies have enough money to fund the research, but not all are doing so. Companies such as British Petroleum are searching for long-term survival by investing in biofuel, solar, and wind power research. Exxon, in contrast, is an example of an oil company that has staked its future on petroleum-based products and is actively ignoring alternative fuel research.[60] This stance apparently makes sense for Exxon because alternative fuels are currently not as profitable as petroleum fuels; not only do the alternatives cost more to produce, but the market for them is much smaller than petroleum fuels.

Will Exxon's strategy pay off? Certainly in the short run. However, should the United States raise gasoline taxes and the price of a barrel of oil rise significantly, the picture may change. As gasoline prices rise, the demand in the United States goes down, at least for a time after the price increase. In that case, Exxon may see its customer base erode over time, along with its profits. And no matter what we do to extend supplies and cut consumption, this planet will eventually run out of petroleum. Considering this last fact, Exxon's strategy begins to seem shortsighted for the long run.

The Best Technology Doesn't Always Win

Sometimes technologies that fill the same niche in our lives reach the market at the same time. If the market isn't big enough to support multiple alternative solutions to the same problem, the market will usually coalesce on one alter-

---

[60]http://money.cnn.com/magazines/fortune/fortune_archive/2007/04/30/8405398/index.htm

native. Probably the best known example of this type of competition between two similar technologies is the Beta versus VHS controversy. Sony released its BetaMax videocassette recorder in the late 1970s; it was followed shortly by JVC's videocassette recorders in the VHS format. The two types of videocassettes were incompatible: Beta tapes could record up to five hours on a tape, VHS, six. Even today, some videophiles believe that the Beta picture quality is superior to that of a VHS tape. Nonetheless, superior technology doesn't always win in the marketplace.

For a time, blank cassettes and recorded movies appeared in both formats equally. However, by the early 1980s the price of a VCR had dropped significantly and VCRs were being purchased in large numbers. VHS machines dropped to as much as $300 below the price of a Beta machine. With lower prices, VHS began to outsell Sony's BetaMax; BetaMax prices also needed to drop but it was too late. By 1984 it was clear that VHS was overtaking the videocassette market. By 1988, Sony had started to manufacture VHS machines, apparently conceding the market to VHS.

A similar battle took place between competing high-definition video formats: HD-DVD (developed by Toshiba) and Blu-ray (developed by Sony). Major companies (entertainment and computer) lined up behind one format or the other. Is one format technologically better than the other? Not necessarily, but they are different and incompatible. They use the same type of laser, but Blu-ray has a higher storage capacity and requires a special coating to protect the contents of the disc. Therefore Blu-ray is more expensive than HD-DVD.

Sony placed Blu-ray players in its PlayStation 3 game consoles, whereas Microsoft backed HD-DVD. Outside the computer field, however, consumers seemed to be holding off, waiting to see which format came out on top.[61] They hadn't forgotten the VHS versus BetaMax battle (especially if they happened to own a BetaMax).[62]

The result of the battle was a win for Blu-ray. Toshiba pulled its HD-DVD product from the market in February 2008. The decision had little to do with the technology itself but instead was market driven. By installing Blu-ray players in the Playstation 3 game console, Sony gained a large installed base without needing to sell separate Blu-ray players. People purchased Blu-ray discs because they had the player in the Playstation 3 and therefore didn't need additional hardware.[63]

Many of the major motion picture studios originally released films in both HD and Blu-ray formats. However, in January 2008 Warner (one of the five biggest studios) decided to release movies only on Blu-ray discs. Given that Sony Pictures, Walt Disney Co., and Twentieth Century Fox were already releasing in Blu-ray only, Warner's actions were just about enough for Toshiba to determine that the HD format wasn't economically viable. The final death knell was the

[61] A quick survey at amazon.com showed 600 Blue-ray products versus 480 HD-DVD products in January 2008.
[62] It is technologically feasible to produce a player that can handle both Blu-ray and HD-DVD discs. At least one company—LG—produced just such a player. (See http://us.lge.com/superblu/?CMP=KNC-BluRay-G) However, it cost more than the combined price of purchasing separate Blu-ray and HD-DVD players.
[63] At the time Toshiba pulled the plug on HD, Blu-ray players sold for around $400; HD-DVD players cost about $150.

decisions of Netflix, the Internet movie rental company, to provide only Blu-ray rentals and Wal-Mart, which sells more CDs and DVDs than any other retailer in the United States, to sell only Blu-ray discs.[64]

The history of the home media marketplace seems to indicate that the market isn't able to support more than one format for entertainment media. Competing formats for CDs and standard DVDs did exist, but global committees decided on standards before producing materials for sale. However, the videocassette and high-definition DVD formats were decided by not only sales figures, but external market factors.

---

[64]For a complete write-up, see http://www.msnbc.msn.com/id/23204819/

WHERE WE'VE BEEN

The North American economy initially was an agricultural economy. As the result of the Industrial Revolution, it became a manufacturing economy. Today we have become primarily a service economy. The change to a service economy has resulted in the loss of some jobs, most notably in manufacturing. However, it has also created jobs to handle the new technologies that are being used.

The rise of the Internet has spawned many businesses. Those that didn't follow good, established business practices tended to fail (like the dot-coms in the early 21st century). Others, backed by patient venture capitalists and with good management, have succeeded. In addition, many traditional brick-and-mortar businesses now have an online presence.

Some businesses—such as banks, publishing houses, and farms—have embraced new technologies and changed as the technologies have changed.

Economics also has effects on innovation. First, innovation must be funded in some way. Second, organizations may try to stifle innovation if it will cause a loss of profits. In addition, when two technologies that are roughly equivalent compete in the marketplace, the winner may not be the technological best but may be determined by market factors.

THINGS TO THINK ABOUT

1. Consider the bank that you use. Explore the bank's Web site and report on the ways in which the bank uses technology. Be sure to pay attention to the measures the bank takes to ensure the identity of the person logging in.

2. As you read in this chapter, the United States has transformed from an industry-based economy to a service-based economy. What risks are there for a country that is primarily a service economy? How can such a country guard against those risks?

3. Besides the three examples discussed in this chapter, identify two other industries that have changed significantly in the past 15 years because of changes in technology. Describe the way in which each has changed.

4. Assume that two companies have offered you jobs. The work is the same for both positions. However, one job is located in the corporate offices, whereas the other is a telecommuting position. If everything else was equal, which would you choose? Why?

5. Would you eat food that came from bioengineered seed or animals that had been fed bioengineered feed? Why or why not?

6. Investigate the negotiations that took place before producers standardized CD and DVD formats. Consider that process and contrast it to the market-driven process that chose VHS over Beta videocassettes and Blu-ray over HD-DVDs. Which is more effective? Should developers of competing formats negotiate a standard or should the marketplace decide which format should prevail? Why?

7. Identify a situation, other than the example in this chapter, where economics has had a negative impact on innovation.

8. Investigate the economics of alternative fuel technologies for cars. At this point in time, which of those technologies is the most cost effective? Which is the most technologically feasible? Which seems to be the most likely to be widely used? Why?

9. Among the roadblocks to a paperless office is the human attachment to documents on paper. Brainstorm and then describe some strategies that a business could use to ease employees from paper-based communication to electronic communication.

WHERE TO GO FROM HERE

Telecommuting

American Telecommuting Association. *http://www.yourata.com/index.html*

Cooney, Michael. "Telecommute. Kill a career?" *http://www.networkworld.com/news/2007/011707-telecommute-career.html*

Gomolski, Barbara. "Confessions of a full-time telecommuter." *http://www.computerworld.com/action/article.do?command= viewArticleBasic&articleId=108966*

Internet Businesses and E-Commerce

Andam, Zorayda Ruth. *e-Commerce and e-Business.* 2003. Full text available for free online at *http://www.apdip.net/publications/iespprimers/eprimer-ecom.pdf*

Battelle, John. *The Search: How Google and its Rivals Rewrote the Rules of Business and Transformed Our Culture.* New York: Portfolio Hardcover, 2005.

Chadbury, Abijit, and Jean-Pierre Kuiboer. *e-Business and e-Commerce Infrastructure.* New York: McGraw-Hill, 2002.

Cohen, Adam. *The Perfect Store: Inside eBay.* Boston: Little, Brown & Company, 2001.

Nissanoff, Daniel. *FutureShop: How the New Auction Culture Will Revolutionize the Way We Buy, Sell and Get the Things We Really Want.* New York: Penguin, 2006.

Seybold, Pat. *Customers.com.* New York: Crown Business Books, 2001.

Spector, Robert. *amazon.com—Get Big Fast: Inside the Revolutionary Business Model That Changed the World.* New York: HarperCollins, 2001.

Vise, David, and Mark Malseed. *The Google Story: Revised Edition.* New York: Delacorte Press, 2008.

Changes in Industry

Barkley, Paul W. "Some Nonfarm Effects of Changes in Agricultural Technology." *http://links.jstor.org/sici?sici=0002–9092(197805)60%3A2%3C309%3ASNEOCI%3E2.0.CO%3B2-J*

Brooke, James. "That Secure Feeling of a Printed Document: The Paperless Office? Not by a Long Shot." *The New York Times,* April 21, 2001.

"eBooks Get Serious." *http://www.idpf.org/pressroom/pressreleases/stats.htm*

Lawton, Krut. "In theYear 2013." *http://farmindustrynews.com/mag/farming_year_3/*

Lyster, Les, and Len Bauer. "The Impact of Technology on Agricultural Extension in the Information Age." *http://www.blackwell-synergy.com/doi/abs/10.1111/j.1744–7976.1995.tb00073.*

Molony, Martin G. "The Effects of Information Technology on Journalism." *http://www.comms.dcu.ie/molonym/maj/*

Muzzi, Doreen. "Agricultural changes only beginning." *http://southeastfarmpress.com/mag/farming_agricultural_changes_beginning/*

Satellite Imaging Corporation. "Agriculture." *http://www.satimagingcorp.com/svc/agriculture.html*

Sucov, Jennifer. "Prepress progression—desktop-publishing trends." *http://findarticles.com/p/articles/mi_m3065/is_n17_v24/ai_17540551*

Williams, Gary W. "Technical Change and Agriculture: Experience of the United States and Implications for Mexico" *http://agrinet.tamu.edu/tamrc/pubs/Im195.htm*

Economics and Innovation

Anthony, Scott O., and Mark Johnson. *Innovator's Guide to Creating New Growth Markets: How to Put Disruptive Innovation to Work.* Boston: Harvard Business School Press, 2008.

"beta vs vhs." *http://tafkac.org/products/beta_vs_vhs.html*

Brown, Warren. "Ponying Up for Alternative Fuel Research." *http://www.washingtonpost.com/wp-dyn/content/article/2006/02/24/AR2006022400778.html*

Clark, Gregory. *A Farewell to Alms: A Brief Economic History of the World.* Princeton, New Jersey: Princeton University Press, 2007.

Dodgson, Mark, David M. Gann, and Ammon Slater. *The Management of Technological Innovation: Strategy and Practice.* Oxford, Oxfordshire, England: Oxford University Press, 2008.

Dubois, David. "The Role of Innovation in Economics." *http://www.infoworld.com/articles/op/xml/02/12/09/0212090pvarian.html?Template=/storypages/ctozone_story.html*

Feldman, Maryann. "The Significance of Innovation." *http://www.competeprosper.ca/images/uploads/Feldman_WIM_Summary_2005.pdf*

Grebb, Michael. "The Showdown: Blu-Ray vs. HD-DVD." *http://forum.ecoustics.com/bbs/messages/34579/129058.html*

Kurzwell, Raymond. "The Economics of Innovation." *http://www.kurzweilai.net/meme/frame.html?main=/articles/art0247.html?*

Laperche, Blandine, and Dimitri Uzunidis, eds. *Powerful Finance and Innovation Trends in a High-Risk Economy*. Basingstoke, Hampshire, England: Palgrave Macmillan, 2008.

Shane, Scott, ed. *The Blackwell Handbook of Technology and Innovation Management* Ames, Iowa: Blackwell Publishers, 2008.

Sperling, Daniel, and Joan Ogden. "The Hope for Hydrogen." *http://www.issues.org/issues/20.3/sperling.html*

Tomlinson, Chris. "Blue Ray VS HD-DVD—Is It Time to Take Sides?" *http://www.waa.co.uk/blog/posts/blue-ray-vs-hddvd-is-it-time-to-take-sides*

Varian, Hal. "The Economics of Innovation." *http://www.infoworld.com/articles/op/xml/02/12/09/0212090pvarian.html?Template=/storypages/ctozone_story.html*

Human Behavior: 7
Communicating and Interacting

WHERE WE'RE GOING

In this chapter we will

- Examine the ways in which technology has changed the way we communicate with one another, including text, audio, and video communications.

- Consider new types of social interactions, such as social networking and virtual communities that technology has made possible.

- Discuss some of the problems that have arisen with new forms of social interactions, including Internet addiction, identity theft, and Internet predators.

- Look at legal measures taken throughout the world to secure the privacy of online data and to define and punish crimes involving electronic communications and data.

INTRODUCTION

During the 1950s a long distance telephone call at home was a major event. It didn't matter what you were doing; you dropped everything and *ran* to get the person for whom the call was intended. Long distance calls were very expensive, so most people didn't make them on a regular basis. Today, we have flat-rate local and long distance plans; we pick up the phone to call anywhere in the North American area code system without thinking about distance or the time we spend on the phone.

It's evident to most people in Western cultures that our social interactions have changed as communications technology has changed. Along with our analog telephone network, we now have electronic communications using the Internet. The Web has given us new opportunities for social interactions as well. In this chapter we will look at the wide variety of changes that technology has made in human society, from our interpersonal communications (probably the most obvious effects) to our family relationships and even to our addictions and mental illnesses.

INTERPERSONAL COMMUNICATIONS

If you were a child growing up in the 1950s and your grandmother sent you a birthday present, you sat down and wrote a thank you letter on paper, folded it into an envelope, put a stamp on it, and put it in the mail. Sending a thank you note is still the polite and responsible thing to do today, but the "letter" may not be written on paper and it may never go in the mail.

Until just a few years ago, salespeople had no choice but to travel thousands of miles to have face-to-face meetings with customers scattered throughout the globe. Today there are technology-based alternatives, in particular, video conferencing.

Do you have a cell phone? If you're like most people living in Western societies, you do. How do you feel if you leave it at home or in your dorm room? Cut off from the world, or relieved that you can't be contacted?

All these technologies—e-mail, video conferencing, and cell phones—mark a major change in the way we communicate with other people. From 1940 onward, there have been a myriad of new communications technologies, some of which have replaced older forms of communication (or, at least, become more important). **Table 7.1** contains a summary of how we have communicated among ourselves, beginning in 1940. This table excludes the type of communication that we have among those we live with, which is primarily face-to-face, but rather is based on how we communicate in the workplace and with those with whom we do not share a dwelling.

Early Communication Technologies

Before the development of communications technology, people used face-to-face meetings and letters to communicate. There were some primitive signaling systems, such as the semaphore system used in England, France, and Germany from approximately 1760 to 1850. It conveyed information through the position of the arms on top of a signaling tower. Semaphore communications were used primarily to exchange military and other official government information.

The first electronic communications technology was the telegraph. Developed in the United States by Samuel Morse in 1837, use of the telegraph was demonstrated in 1844. The telegraph became so successful that a transatlantic cable was in place by 1858; a transpacific cable was laid in 1902.

The introduction of the telegraph, along with the railroads, brought the United States, and ultimately, the world, closer together. Although sending a telegraph message was expensive, it was the first technology that allowed close to instantaneous communications. Before the telegraph it could takes weeks for the results of a presidential election to be disseminated across the United States; the telegraph let the entire country know in a few minutes.[1]

---

[1] The United States tried another method of fast communications before the telegraph: the Pony Express. The idea was to use a relay of riders on fast horses to carry the mail. Only small volumes of mail could be sent this way: It all had to fit in the saddlebags of a single horse and couldn't be heavy enough to slow down the animal. The Pony Express was put out of business by the introduction of the telegraph.

Table 7.1 How We Communicate

Era	Modes of Interpersonal Communication (listed in order of importance)
1940–1960	Face-to-face meetings Letters Local telephone Long distance telephone (not common for home users because it was very costly) Telegram
1960–1980	Face-to-face meetings Local and long distance telephone (long distance rates beginning to come down) Letters E-mail Telegram
1980–2000	Local and long distance telephone Cell phones E-mail Face-to-face meetings Video conferences Letters Internet telephony Telegram*
Beyond 2000	Local and long distance telephone E-mail Spoken cell phone conversations Text messaging Face-to-face meetings Internet telephony Video conferences Letters

* Western Union discontinued telegrams in 2001, although hand delivery stopped in 1972. See http://www.retro-gram.com/telegramhistory.html for a history of telegrams.

The Morse code used by telegraphy was clumsy. As you can see in **Table 7.2**, Morse is a binary code (made up of two states), but the codes are of different lengths. Telegraphy required both a human sender and a human receiver. The sender tapped the symbols to be sent on a key (**Figure 7.1**), leaving different time gaps between letters, words, and sentences.[2] The receiver had to interpret the long and short clicks produced by this key, along with the differing lengths of

---

[2] Morse's letter patterns weren't arbitrary. As you can see in Table 7.2, the more frequently a letter is used, the shorter its code.

Table 7.2 Morse Code Used to Send a Telegraph

Symbol	Code Sequence	Symbol	Code Sequence
A	•—	S	•••
B	—•••	T	—
C	—•—•	U	••—
D	—••	V	•••—
E	•	W	•——
F	••—•	X	—•—•
G	——•	Y	—•——
H	••••	Z	——••
I	••	1	•————
J	•———	2	••———
K	—•—	3	•••——
L	——••	4	••••—
M	——	5	•••••
N	—•	6	—••••
O	———	7	——•••
P	•——•	8	———••
Q	——•—	9	————•
R	•—•	0	—————

silence between characters, words, and sentences. Because of this, most telegraph messages were short; they weren't used for everyday business transactions but for extraordinary events.

Telegraph messages were used to announce major events. Usually, all that could be sent was the equivalent of a headline; details often followed weeks later in a newspaper or letter. Families could be notified of births and deaths; remote towns could be notified about political events, the progress of a war, or natural disasters. People who lived far apart began to feel more connected. In fact, you could postulate that the telegraph was instrumental in getting the United States to feel like a single country after the Civil War because it allowed fast communications.

Voice Communications

Technological communications didn't become practical for day-to-day use until the introduction of the telephone. Before 1980 in the United States, most people had telephone service provided by one company: AT&T. There were—and still are—some small, independent, local providers (for example, Western Electric that still continues to service areas in Multnomah County, Oregon that are outside the border of the city of Portland).

Figure 7.1
A typical telegraph key.

Local telephone rates were kept low by subsidies from high long distance rates. As mentioned earlier in this chapter, long distance rates were so high that long distance calls were not made without some forethought. Therefore it was relatively cheap to call across town (or, at least, within your local calling area), but using a long distance call to talk with friends and relatives far away was a special occasion.

Early home telephones in the United States were often connected to a "party line," where multiple phones shared the same line. Each phone had a separate ring, but anyone could pick up a phone and hear whatever conversation was on the line at the time. Listening to someone else's conversation was known as "rubbering" and became a favorite pastime with those who liked gossip or who had nothing better to do. Rubbering was considered very bad manners, but that certainly didn't keep it from occurring. Today, party lines no longer exist in the United States, and rubbering as a human behavior has been eliminated (outside of picking up an extension phone).

European telephone systems primarily were owned by governments.[3] Local service usually was not subsidized by long distance rates and therefore was more expensive than in the United States. All local calls were charge using "message units," where calls were priced based on the length of the call. (U.S. users could choose a message unit service or a flat rate local service, but most European users had no choice.[4]) The cost of phone service meant that even in the 1970s many homes did not have telephones.

Long distance telephone use didn't become an everyday occurrence for individuals in the United States until the breakup of AT&T in 1980. By that time, a few independent long distance service providers were in business (in particular, MCI and Sprint), significantly lowering the cost of long distance calls, but users

---

[3] People used to wonder how the English could have such a good mail system—delivery from one end of the country to the other in a day and two deliveries each day—and yet have such a crazy phone system. Each town had its own dialing prefix, but unlike a North American area code, the prefix was different depending on where you were calling from. Each location had its own little red directory for the calling codes from the local exchange. Only the large cities had the same prefix from any place in the country.
[4] It was common courtesy in England, for example, to leave the cost of a call by the phone when you made a call from someone else's house.

needed to dial a long code to gain access to an alternative network from an AT&T local loop. The breakup of AT&T separated AT&T's long distance business from its local telephone business. The latter was to be provided by regional companies (the "Baby Bells"), some of which have reconsolidated today. It was at that time that customers were allowed to choose any long distance provider and access that provider's lines without the dial-around code.

The intent of the breakup of the AT&T monopoly was to stimulate competition and ultimately bring down telephone rates. Long distance charges did come down—significantly—but local charges, which were no longer subsidized by long distance revenues, went up. Initially, most users didn't see much of a change in their total phone bill.

Landline telephone bills didn't start to decline noticeably until the telephone companies began to feel pressure from cell phones. It is becoming more and more common for people not to have a landline phone at all but to rely solely on a cell phone. There are some interesting ramifications of this. First, a person has one phone number that he or she can theoretically keep for life.[5] Second, you may not know where you are calling when you dial a cell phone number. We used to know where we were calling based on a phone number's area code and take time zones into consideration. But with the mobility of a cell phone, there is no way to know whether you will be waking someone out of a deep sleep when you call.

Cell phones have also brought some new social rules with them. In most Western cultures it's considered impolite to talk on the telephone in a restaurant. If you go to a live performance or a movie, it's polite to turn off your cell phone (or at least set the phone to vibrate only when someone calls).

Many people like their cell phones because the phones make them always accessible. Others, however, prefer to turn off their phones so they spend some time away from others. Cell phones have made either choice possible.

Voice communications are now also available using the Internet. Voice Over IP (VoIP) uses a broadband Internet connection to transmit voice, bypassing standard telephone wiring altogether. At this point in time, VoIP is the cheapest way to get telephone service. However, there are two issues that keep it from being widely adopted. First, broadband access is limited in most countries. Second, if the Internet becomes inaccessible for any reason, VoIP doesn't work. If you aren't comfortable with being without phone service, you'll need a backup cell phone, probably eliminating any cost savings.

The telephone—landline, cell, or VoIP—does cut down on face-to-face interactions. Rather than going across town to talk to someone, you just pick up the phone; rather than going to a store to see if a product is in stock, you make a phone call.

The introduction of the telephone also cut down on written communications. Rather than sending a letter to someone to bring that person up to date on your life, you make a phone call. Rather than sending an order to a catalog company, you can place the order over the phone.

---

[5] The ability to keep your cell phone number when you switch cell phone providers is fairly recent and you may occasionally still run into a provider who is unable to transfer an existing number.

Telephones also can be socially disruptive. How easy is it to ignore a ringing phone? For most of us, not too easy at all. We'll drop just about anything to get that phone call. It doesn't matter what else we're doing, we *have* to answer that phone!

It didn't take retailers long after the dispersion of the phone across the United States to discover that they could make calls to potential customers to advertise their products. Telemarketing calls are at best informative, but most people consider them intrusive, especially because marketers have a habit of calling at the dinner hour. Some of the marketing calls are also intended to defraud, cadging money from segments of society (for example, seniors) who are most likely to donate to what they think is a worthy cause or to purchase products they don't need.[6]

Our perceptions and acceptance of the telephone have changed over time. Initially, people found the telephone to be disruptive and many weren't eager to have one in their homes. It took a generation for the telephone to become an accepted part of our lives. A similar reaction greeted the telephone answering machine. At first, people would complain: "I don't want to talk to a machine." Now that such devices are commonplace, we are annoyed if someone doesn't have one so we can leave a message when we call. We've also discovered that answering machines can be used to screen calls, and sometimes it is easier to leave a message on a machine than it is to talk to someone, especially if the topic is embarrassing or emotionally difficult.

Another telephone technology that initially met with some skepticism was caller identification, or caller ID. Some people believed it was an invasion of privacy. The telephone companies responded to these concerns by implementing the technology to block a telephone number from appearing on a caller ID screen.[7]

Today most people in developed societies are never far from a telephone. If you don't have a cell phone on you, it's highly likely that the person next to you will. And we use those cell phones a lot. Research done by my students has shown that during a class break, around 30% of the students walking on campus are using a cell phone (either looking at it, dialing, answering, or talking).

Video Communications

Another one of those predictions for the future from the 1950s was the video telephone. A two-way video phone would transmit real-time video from one phone to the other, using telephone lines. There have been several attempts to implement video phones, but the wired telephone system just doesn't have the bandwidth to stream video. In addition, some people were leery of the idea of video phones: They didn't want to be seen by someone else without careful preparation (no curlers, or nightgowns, or torn T-shirts).

Early video conferencing involved the broadcast of a television signal, usually via satellite; audio wasn't necessarily included but might be provided by telephone. To participate in a video conference you often had to travel to a special location where the television camera and satellite uplink/downlink facilities were available.

---

[6] In the United States, the "no call" registry has helped stop unwanted solicitations However, political and charitable organizations are exempt, as are businesses with which you have an ongoing relationship.

[7] Blocking your telephone number doesn't always work. For example, when you call the telephone company, they know the number from which you're calling, even if you normally block every call. Other businesses, such as some credit card issuing banks, bypass the caller ID blocks as well.

The equivalent of day-to-day video communications has waited until the Internet could handle the video. Video chats, with a camera at either end of the conversation, have filled the niche originally postulated for video phones.

For the most part, video conferencing is a good thing. However, like any other technology, taking it to its extreme could have unwanted consequences. Isaac Asimov, the science fiction author, wrote about just such a society in his novel, *The Naked Sun*. Part of his Robots sequence, the story involves humans who had become so isolated that they couldn't stand to be in the presence of other humans but instead used video communications to interact with their neighbors. The necessary tasks of daily living were performed by robots.[8]

Text Communications

Text communications come in two major forms today: e-mail and text messaging. Blogs ("Web logs"), instant messages, and chat can also be considered forms of text communication. Because text communications of all kinds prevents from seeing the person at the other end of the conversation, the Internet is sometimes known as the "great leveler." It can hide differences in age, race, size, and other physical attributes that might interfere with the forming of friendships (or other relationships). This is not necessarily a bad thing. When we can't react to the physical presence or even the voice of a partner in a conversation, we have to focus on what a person is saying and how it is being said. Language and ideas become more important than physical appearance.

E–Mail Originally, people wrote letters. A "pen pal" was someone with whom you exchanged handwritten missives. But how many letters have you sent in the past year? Perhaps a thank you note to a relative for a gift or an R.S.V.P. to a wedding, using a reply card and preaddressed, stamped envelope. Perhaps you've paid a bill or two through the mail using a check or money order. More and more, however, many of us have turned away from paper-based communications and toward electronic communications.

Virtually everyone who has access to the Internet at home or at work uses e-mail. It has replaced phone calls (both individual and conference), face-to-face meetings, and paper memos. E-mails sent from some calendar applications to announce a meeting will automatically place the meeting on the recipient's electronic calendar. E-mails can also contain still pictures, video, and audio, depending on the features provided by your e-mail application.

Why do people like e-mail so much? It's quick, it's easy, and it's always available. If the person you call on the telephone doesn't answer, the best you can hope for is to be able to leave a message on an answering machine. However, we generally consider answering machines to be public, and people are often reluctant to leave private messages that anyone can hear. E-mail, in contrast, can be read when the recipient returns to his or her computer, and the message is relatively private.

---

[8] Asimov never addresses the question of how these people got together to have children, or how children would be raised. The scenario he spins does make one wonder, however.

E-mail messages tend to be shorter than written letters and a bit more abrupt. For example, when writing a personal letter you begin with something such as "Dear so-and-so" and close with "Sincerely" or "Love" followed by your name. E-mail, however, usually starts with just the name of the recipient or no name at all; the sender just starts typing the message. Most people are used to the style of e-mail and are not offended when e-mail doesn't completely adhere to the style of a letter.

Some people view the more abrupt style of e-mail as a degradation of a more civilized form of communication (the letter). However, anecdotal evidence suggests that this attitude is held by those who did not grow up with e-mail and may therefore disappear in a generation.

Text Messaging Text messaging sends text-only messages via cell phone. The younger you are, the more likely you are to use text messaging. Because the keyboards on cell phones are small and three or more characters are tied to each button, typing text on the cell phone is clumsy. Users therefore have developed a shorthand for use in electronic communications. For the most part, it's a phonetic version of English, with some abbreviations thrown in: r u coming? u r? thx. Translation: Are you coming? You are? Thanks.

There is major concern among parents and educators that the text messaging shorthand will carry over into more formal communications, such as papers that are written for school. However, this doesn't appear to be the case. Young English speakers seem to have as many as four versions of the language: the one they use for text messaging (and also instant messages and chat), a spoken language for use with friends, a spoken language for use with adults, and a formal written language. They switch between the versions of English almost effortlessly, depending on the situation.[9]

Instant Messages and Chat Instant messaging (IM) first appeared as long ago as the 1970s, but it didn't enter common use until it was implemented by America Online (AOL). Initially, AOL Instant Messenger (AIM) was accessible only by to AOL customers. However, today client software is available for the major computing platforms. Other vendors, such as Microsoft, also offer IM software. For the most part, instant messaging is free.

Instant messages often replace phone calls. Because they use the Internet, they can cost much less than telephoning, especially for overseas communications. It is also possible to use IMs to carry on multiple conversations at the same time, something you can't do with a telephone.

In some cases IMs have replaced face-to-face communications. Assume that you're a college student living in a dorm. Before IMs, you would walk down the hall if you needed to talk to someone. Today, however, you would be just as likely to IM someone, even if that person was in close physical proximity.

---

[9] Using the wrong form of the language can have dire consequences. For example, if a teen were to use "teenspeak" with his or her parent, the parent would most likely see the speech as disrespectful. If you were to use text-messaging shorthand in a graded paper for class, the grade would suffer significantly.

Chat is much like an instant message but usually involves more than two people. In its basic form, chat is text only. However, some computers do support video chat, using a camera (even a digital video camera will do) and a fast Internet connection.

Blogs Blogs have become increasing popular and important in the past few years. Originally intended as public, online diaries, they have become important venues for the sharing of information and the gathering of opinions. Blogs can cover legal and philosophical issues, such as a blog dealing with copyright and fair use laws.[10] They can also be used by a family to share information with friends and relatives, creating an always-changing electronic form of the "Christmas letter."[11]

There was nothing equivalent to a blog in the pre-Internet era. There was no way to provide such widely viewed exchanges of information in a timely manner. Some groups of scholars used a "round robin" technique of exchanging letters that moved from one person to another in a circle, but such exchanges were slow and could not be viewed by the entire group at the time same. Newspapers disseminate information relatively quickly, but responses to articles—letters to the editor—usually take at least a week after the original article to appear. In addition, not all letters to the editor are published; the paper's editorial staff actively chooses the letters to print. Blog postings, however, are much more open; just about anyone can post to a blog, although the blog's owner usually can remove unwanted entries.

Anyone can start a blog. Many ISPs now provide blogging[12] capabilities along with e-mail accounts. However, there is no one to vet millions of blogs on the Internet to determine which are valuable to any given individual. The biggest question that arises is whether the information in a given blog is authentic and authoritative. Because no one regulates what is posted on the Internet, it is up to each individual user to decide.[13,14]

Wikis A "wiki" is a collaborative piece of online writing. The first wikis were established in the mid-1990s. The most well known is probably Wikipedia, the Web-based encyclopedia that began operation in 2001.[15] In concept, anyone can write an article for Wikipedia; anyone else can edit that article. The idea is something similar to "regression toward the mean," a statistical concept through which extreme values tend to modify over time and move toward the mean (the arithmetic average). In practice, this means that as an article is edited, it becomes increasingly more accurate, better written, and better documented.

---

[10] See http://thefairuseblog.typepad.com/

[11] For examples, see http://thebayerfamily.blogspot.com/, http://reyesblog.dabu.com/, or http://www.tongfamily.com/

[12] "Blogging" joins "googling" and "scrapbooking" as words that have been coined from nouns to describe the process of using or creating something.

[13] As noted earlier in this book, some countries do filter what their citizens can access. They may also stop individuals from posting opinions that are antigovernment. However, as a whole, the Internet is open to just about anything anyone cares to post. The only true taboo seems to be child pornography, and even that isn't totally absent.

[14] Blogs can spread gossip, rumors, and innuendo. See Gossip, Reputation, and Shaming on the Internet, later in this chapter.

[15] http://www.wikipedia.org

As of the time this book was written, Wikipedia had more than 2.3 million articles in English and hundreds of thousands of articles in languages such as Dutch, Spanish, French, Italian, German, Portuguese, Norwegian, Romanian, Finnish, Turkish, Arabic, Greek, Danish, Russian, and Swedish.[16] The total number of articles in all languages reached a total of 10 million in March 2008, making it the largest encyclopedia in the world.

Anyone with the necessary software and a Web server can establish a wiki. They can be dedicated to any project for which collaborative editing is appropriate. There is no reason, for example, why a business couldn't use a wiki to allow many employees to contribute to a corporate document. Researchers are also using wikis to make it easier for them to share ideas with others working in the same field. Although wiki articles tend to gravitate to a consensus of what is correct as they are modified over time, their very nature means they vary in quality.[17]

To many people, wikis represent the essence of the Internet. They are a collaborative, free medium in which everyone can participate. Like the open source software movement, they can tap the collective expertise of the world. However, that very openness makes wikis vulnerable to inaccuracies, poor writing, copyright infringement, and contributors with strong biases. Most Wikipedia articles are open to anonymous editing by anyone, for example. However, some articles have proven to be controversial, subject to what Wikipedia calls "edit wars," in which two or more authors continuously delete another's contributions and replace it with their own; users who object to a specific article have also been known to deface the article (cyber-vandalism). Wikipedia has therefore restricted editing of those pages by requiring users to have confirmed accounts.[18]

Junk Mail, Spam, and Associated Garbage

Beginning in September each year, many North American mailboxes are weighed down by holiday catalogs, many from companies of which the recipient has never heard. This type of advertising, which increases in volume before the Christmas buying season, puts hundreds of unsolicited mail items in mailboxes. We call it junk mail, and most of the time we just toss it in the trash.[19] What bothers us about junk mail isn't that we receive unsolicited mail; we've gotten used to it and most of us have no problem ignoring it. Rather, people today are more concerned about the waste of paper and the energy needed to print, bind, and mail the advertising.

The situation with e-mail is similar in the sense that most of us receive unsolicited e-mails.[20] However, e-mail junk, known as "spam," is potentially more

---

[16] Wikipedia articles are written in 250 languages.

[17] Because it is an encyclopedia, many instructors will not allow students to use Wikipedia articles in their schoolwork. However, many Wikipedia articles are documented with links to external Web sites and citations to print materials. The reference lists can therefore act as a springboard to research.

[18] See http://en.wikipedia.org/wiki/Wikipedia:Protected_page for information on Wikipedia's protection policy.

[19] Much of the advertising by mail can be stopped by adding yourself to a mail preferences list, such as the Direct Marketing Association's Mail Preference Service (www.privacyrights.org). There is a $1 fee for including your address on the list.

[20] The only way to prevent unwanted e-mails is not to use the e-mail address on any Web site. Don't register for anything; don't purchase anything. You may therefore want to maintain at least two e-mail accounts, one that you use for registering with Web sites and making online purchases and a second that you use for regular correspondence. This won't work, however, if you have a "friend" who shares your private e-mail address with other friends and mailing lists.

harmful than unwanted advertising that comes in the mail. At the very least, it costs time to sort out the junk mail. Even more seriously, e-mail is used to deliver scams and "malware" (viruses and other nasty bits of unwanted software). The scams range from fraudulent investment schemes to "phishing," where a person is tricked into visiting a Web site where he or she enters personal financial information that can be used for identity theft.

It can be very difficult to recognize a phishing attempt; the e-mails purport to come from a well-known site such as eBay or PayPal and use graphics from the legitimate Web site.[21] They play on human insecurities by threatening to close down the recipient's account if he or she doesn't update or validate account information. The spoofing Web site asks for every bit of financial and identifying information that anyone might need.[22]

Viruses have been distributed via the Internet since at least 1980. Originally, they were mildly annoying pranks, software that displayed funny messages on the screen. However, the pranks turned serious, with viruses that destroyed the contents of hard disks or tied up a corporate network with so much traffic that legitimate users couldn't send messages.

Today, malware consists of threats in addition to viruses, including worms, "spyware," and "bots." Bots are particularly nasty. A bot is installed on a computer without the user's knowledge and may sit idle for a long time. However, when the controller needs computing power, he or she sends a command to the bot to use the infected computer for whatever task the bot was programmed to perform, turning the machine into a "zombie." Thousands of zombies working together, even though scattered throughout the world, can provide a significant amount of computing power.[23]

SOCIAL NETWORKING

Most forms of communication discussed in the preceding section didn't arise with the Internet; they've been used far longer than that. However, the World Wide Web has provided a new form of communication, which has come to be called "social networking." Two of the most well known sites are MySpace[24] and Facebook.[25] The general purpose of each site is to provide a place for people to introduce themselves to the virtual community and to form relationships between individuals. The two sites do differ in the audience each has been designed to serve.

MySpace has always been available to the general public. The basis of MySpace is your personal profile and photos. You can also contact other users through e-mail and instant messages as well as maintain your own blog. One of the ideas

---

[21] See http://www.about-the-web.com/shtml/scams.shtml for a list of current e-mail scams.
[22] Phishing to gather information for identity theft isn't limited to e-mail. One particularly insidious scam in the United States involves a phone call during which the caller says he or she is from the IRS. The person called has forgotten to put a social security number on a recent tax form. Would the called person let the caller know what it is so the tax form can be straightened out? U.S. citizens fear tax audits, and most are happy to give the information just to clear things with the IRS. Known by the general term of "social engineering," this type of scam works because it plays off existing fears and insecurities.
[23] It was just such a bot network that perpetrated the attack on Estonia described earlier in this book.
[24] www.myspace.com
[25] www.facebook.com

behind MySpace is that you can extend your circle of friends easily. You make a list of MySpace friends for your own use. In doing so, you are connected automatically to all your friend's friends. This makes it very easy to amass a large network of contacts with which you can share information.

MySpace lets a person share just about any personal information he or she wants to make public. The simplest way is to use MySpace's profile pages. You can also use the blog features to post any musings that one might have about life, the universe, and anything else.

At first glance, Facebook seems to be the same as MySpace. However, it has a slightly different history and different behaviors when it comes to connecting people. Facebook began as a site for high school and college students; it was opened to the general public only recently. Facebook's friend lists don't cascade like those of MySpace: When you add a friend to a list, that one person is all that is added. You choose exactly whom to add.

In 2007 both Facebook and MySpace began using the data submitted by members for other purposes. MySpace created "hypertargeting," a collation of information from user profiles that allow retailers to better target their marketing.[26] Facebook's Beacon is software that shares a user's actions at online retailers—purchasing, registering, adding an item to a wish list—with all their Facebook friends.[27] There are some major social issues associated with these activities. Probably the most benign effect is purchasing a gift for someone on your Facebook friends list—he or she will find out what you bought. More importantly, both hypertargeting and Beacon provide benefits to retailers without benefiting users; marketers gain at the loss of users's privacy. Users do know that when they join a social networking site they will receive advertising, and neither MySpace nor Facebook is doing anything illegal, but to many people these actions seem like an unwarranted invasion of privacy.

VIRTUAL COMMUNITIES

A virtual community is a group of people who interact with one another as if they were in a face-to-face social situation. One of the largest and most well known is Second Life.[28] Each person has an avatar, an online representation of the human. The avatar begins with a generic look and shape. However, the first thing a new user probably wants to do is change the avatar's appearance in some way.

Second Life provides a three-dimensional visual landscape (or as three-dimensional as a standard computer monitor can be). The landscape changes as an avatar moves, providing access to a vast, self-contained world. Although virtual worlds such as Second Life began as social enterprises, allowing avatars to interact with one another using text or audio chat, many virtual worlds have evolved in locations where users can get information, purchase virtual real estate (and build dwellings that can then be furnished), and purchase items just as they might from any commercial Web site.

---

[26] See http://biz.yahoo.com/bw/071105/20071105005655.html?.v=1
[27] See http://www.facebook.com/business/?beacon
[28] www.secondlife.com

Although basic entry and use of a virtual community is typically free, such Web-based applications are commercial concerns. Many items with which you can equip your avatar cost virtual dollars, which are purchased with real-world funds, In addition, the free use of the virtual world is supported by a great deal of advertising. For example, as you move through Second Life you encounter a myriad of billboards that advertise real-world businesses.

INTERPERSONAL RELATIONSHIPS

It's not unusual to form online friendships with people with whom you communicate frequently. Sometimes these friendships grow into romantic relationships. The Internet makes it possible to meet people who are far from you, people whom you would otherwise never have a chance to know. You have a chance to meet others with similar interests and concerns who are geographically scattered. Most of us would consider this to be a benefit of communications technology.

Occasionally, online friends want to meet in the real world. If the parties involved have been honest with one another, then the face-to-face meeting will probably go well and a new, deeper friendship that might never have happened without the Internet is cemented. However, if the parties have misrepresented themselves, then the situation can become very dangerous. (See Are You Who You Say You Are?, later in this chapter for more on this subject.)

Dating and matchmaking agencies have been around for many years. The ability of the Internet to connect people who are far apart makes it an ideal vehicle for running a data service. Dating sites usually require you to fill out a profile form with information both about yourself and about what you are looking for in a partner. Then they match your information against others in their database. The cost often depends on the number of matches you want to purchase; you might, for example, receive five matches for your initial sign-up fee and then be required to pay for additional matches. Just how good the matches will be depends, as you might expect, on who else is in the data service's database.

The Internet has also spawned Web sites for companies that act as "marriage brokers." Exactly what these sites do—and whether they are legitimate—varies a great deal. One inexpensive site takes your profile and matches it against the many dating sites already on the Internet. All it does is save you the time of searching existing electronic dating material already on the Internet.

Some sites (many coming from the Ukraine and Russia) offer to match American men with Ukrainian/Russian women and ask for money to send the woman to the United States to meet the man. In some cases, the women simply do not exist; in other cases, the request for money is fraudulent because U.S. entry visas cannot be purchased. Although there are legitimate Internet-based marriage brokers that do match American men and Ukrainian/Russian women, the men usually must travel to the Ukraine or Russia to meet the women. Unfortunately, many of these agencies are scams.

FAMILY RELATIONSHIPS

Technology has also changed family relationships. In particular, the multigenerational household is a rarity today. There are several reasons why this is so:

- With the move away from a farming economy, multigenerational families aren't needed to make a living.
- Transportation technology—motor vehicles, railroads, airplanes—allow families to spread out because it is easy to travel to be with one another.
- Communications technology makes it easy to stay in touch with family, even if the family members are thousands of miles apart.

As well as allowing families to live further apart, technology has made it possible for some people to be families who couldn't before using assisted reproduction technologies. Although we will talk more about reproductive technology in Chapter 10, keep in mind that before the development of in vitro fertilization, childless couples that wanted children could adopt or remain childless. Technology has allowed them to become parents to their own biological children.[29] By the same token, medical technology has given women control over their fertility through the birth control pills and other birth control devices. At no time in history has it been easier for women who don't want children to stay child free. Assisted reproduction isn't easy—the success rates aren't very high and the procedures are very costly—but the technology works for many. Technology has given us choices as to when we have children and how many children we have.

SOCIAL ROLES: THE EMERGENCE OF GEEKS AND NERDS

The focus of our society on technology has brought two groups of people to public awareness: geeks and nerds. Both groups include those who spend most of their intellectual and social lives on technology and both groups are generally considered to contain individuals who are outside mainstream Western social norms.

Although the terms "geek" and "nerd" are sometimes used interchangeably, they are subtly different. A nerd is usually someone who is focused on technology to the exclusion of other academic and social interests. The typical picture of a nerd is someone who knows little about personal grooming (the kid with the pants halfway up his chest and tape holding his glasses together) and who has few social graces. Nerds are often considered to be socially irritating and not fit for company beyond other nerds. Nerds appear to be unaware of their social ineptitude. Despite their social problems, nerds are usually kind and nonviolent. In general, calling someone a nerd is an insult, although the nerd himself may not recognize it as such.

---

[29] Certainly we can argue as to whether the drive to have a child that is biologically related to you is important. We can also acknowledge that it is important to some people. These are the individuals who choose either to remain childless or attempt conception using reproduction technologies.

Geeks are computer experts who often wear the label geek with pride, especially when they apply the term to themselves. They are not necessarily as socially unacceptable as nerds, although they may find it difficult to talk about anything other than computers in a social setting. In fact, a computer repair company in the United States has adopted the term, calling themselves the Geek Squad.[30] Nonetheless, a geek is considered to be more focused on technology than the typical person in Western society.

Nerds and geeks didn't spring from computing technology: They've just become more noticeable. Nerds have been around for some time. In the 1950s, for example, nerds wore slide rules on their belts and dressed much like nerds do today.[31] In the 1960s the slide rule was replaced by the pocket calculator. Nerds could also often be recognized by the presence of a pocket protector in a shirt pocket. Before the rise of computing, the most recognizable nerds were engineers.

Some nerds suffer from Asperger syndrome, a disorder on the autism spectrum.[32] Those with Asperger syndrome are often highly intelligent but have significant communication and social problems. They may also have gross motor developmental delays, making them clumsy and ineffective at sports. A single area of expertise (be it computers or mathematics or any other discipline) is also characteristic of the disorder. Interacting with computers is easier for many Asperger syndrome sufferers than interacting with people. They find jobs such as programming, which is largely a solitary activity, easier to manage than jobs that require more social contact. Asperger syndrome, however, is not a new disorder; it has been a part of the human condition for a very long time—it isn't something that sprung up as a result of technological development—but we have recognized it medically only recently and are still learning the best way to treat individuals who suffer from it.[33] The question remains, nonetheless, as to whether Asperger syndrome sufferers need treatment or whether society as a whole should accept individuals who are different in this way.

Geeks seem to be a newer phenomenon than nerds, becoming generally recognized with the rise of computers. The term "geek," however, has been in use far longer than computers have existed. Its original meaning was someone who was a freak, such as a sideshow performer. The irony of this is that today nerds are usually considered more freakish than geeks.

PRIVACY

We hear a lot today about the security of electronically stored information (or the lack thereof) and much less about privacy. The two aren't the same: Privacy is the need to keep specific information from disclosure to unauthorized people; security is what we do to keep private things private.

---

[30] Geek Squad is the computer service arm of Best Buy. You can find their official site at http://www.geeksquad.com/?PSRCH. Technicians drive black and white Volkswagen Beetles with the company logo painted on the sides; they are easily identifiable on the streets.

[31] The stereotypical nerd uniform includes a short-sleeved white dress shirt.

[32] For more information on Asperger syndrome see http://www.aspergers.com/

[33] For a discussion of the psychology of nerds, see http://www.mentalhelp.net/poc/view_doc.php?type=advice&id=2579&at=2&cn=91&ad_2=1&submit=I+Agree

Electronic data storage has affected our privacy in profound ways. Even before there were databases of personal information stored on millions of computers, however, there were public sources to information we often consider private. For example, many states have traditionally sold driver's license data to firms that wanted to send advertising to licensed drivers. Property ownership data have been available at town halls since colonial times; anyone can go into the town clerk's office and look at the information, including the names and phone numbers of property owners, even if the phone numbers are unlisted. In fact, public information includes all the following:

- Property tax information, including assessments, tax liens, and judgments against property owners for non-payment of taxes
- Motor vehicle records, both car registrations and driver's licenses
- Voter registrations[34]
- Business and professional licenses
- Bankruptcy filings
- Criminal warrants, arrests, and convictions
- The results of civil court actions

The advent of computer-based databases has meant that such information can be disseminated more easily (and at much less expense). It also meant that more information could be stored. There is a great deal of concern today about the privacy of personal information. However, personal information has always been available; it just hasn't been as easy to obtain as it is when the data are stored electronically.

The most widely known issue we face today because of a lack of privacy is identity theft. If a thief can obtain a social security number, a credit card number, and checking account information, then he or she can get credit in another name and even request a social security card. With that social security card, the thief can obtain a driver's license and a birth certificate and a passport. Although you are probably not liable for fraudulent charges made with your credit cards, erasing other damage an identity thief has done can be frustrating, difficult, and sometimes even impossible.

Identity theft has an impact on business as well as on individuals. Summit Electric Supply, a company based in Albuquerque, New Mexico, was the victim of an identity theft scam.[35] A thief represented himself as being an agent of the company and ordered hard disks and other electronic components. Most wholesalers ship without requiring payment up front. The thief took delivery of the merchandise and, by the time the bill arrived, was long gone, off somewhere selling the illegally obtained components.

Stopping identity theft is difficult. It means that a wholesaler or retailer needs to authenticate a person attempting to make a purchase with a check or using credit. In a corporate world, authentication could be as simple as taking the ordering company's name, looking up the company's phone number, and calling back to verify that the order is really coming from the company. With people and credit cards it's more difficult, because an identity thief usually isn't using a stolen credit card but a legal credit card obtained using a stolen social security number. Authentication of credit cards using the three-digit number on the back of the card only ensures that the card is in the hands of the person doing the ordering.

You need three things to positively identify a person: something the person has, something the person knows, and something the person is. Something the person has could be a credit card or an account name. Something a person knows is a password. Something a person "is" involves a biometric identifier, such as a fingerprint or a face scan. We do very well with the first two elements of identification, but the technology of biometrics is still in its infancy. The most typical biometrics today are fingerprint scanners, some of which are built into laptops; others are simple USB devices.

GOSSIP, REPUTATION, AND SHAMING ON THE INTERNET

The unregulated nature of the Internet makes it an ideal place to spread gossip and other unsubstantiated facts about people. It an also be used to expose unacceptable behavior and punish the transgressor through shaming. Before the Internet, the sharing of information that could affect a person's reputation was typically limited to the person's social circle, a relatively small group of people. In addition, damage to a person's reputation, whether caused by shaming for an actual transgression or inaccurate gossip, would repair itself over time. Both of these things are not necessarily true when the Internet is involved.

Once something has been posted to a publically accessible Web site such as a blog, there is virtually no way to prevent others from taking the information from the site (including any illustrations that might be there) and posting the same material on yet another site. Rumors and reports of inappropriate behavior can spread like wildfire. Even if the original posting is removed, it is often impossible to stop the spread of such information.

How bad can it be? In 2005, a young Korean woman was riding on a subway train with a small dog. The dog eliminated on the train floor, and the girl refused to clean up the mess, even when told to do so by those around her on the train. Another passenger with a camera cell phone took pictures of the incident and posted them on the Internet. It took only a few days for the young woman's identity to accompany the pictures that were spreading from one Web site to another. Koreans found her behavior reprehensible and used blogs to state their opinions forcefully. The result was that the young woman (given a nickname by bloggers that was the Korean equivalent of "dog poop girl") lost all privacy in her life. She was recognized on the streets and eventually lost her job.[36]

When the young woman failed to clean up after her dog, she violated what was considered appropriate behavior. People who were not present during the incident took

---

[34] Some states do restrict access to voter registration information.

[35] For the entire article describing what happened to Summit Electric, see http://www.crn.com/it-channel/202404442

[36] More information and opinions of this incident can be found at http://www.docuverse.com:80/blog/-/alias/donpark/e5e366f9-050f-4901-98d2-b4d26bedc3e1 and http://www.washingtonpost.com/wp-dyn/content/article/2005/07/06/AR2005070601953.html

what they read on the Internet and strongly condemned her. However, in this case the punishment probably did not fit the crime. Although the young woman's behavior was tacky at best, most people would agree that she didn't deserve to lose all privacy and her job; her offense wasn't worth having her life ruined. Given the open nature of the Internet, however, currently there is no way to stop the spread of such information.

This is not to say that all reporting of personal transgressions on the Internet is negative. When an inappropriate behavior is revealed to others, the transgressor feels ashamed. Shaming has long been used as a punishment for those who violate societal behavior codes. For example, the Puritans placed offenders in the stocks. Shaming was used for those acts that offended what the community believed was "decent." It was a way to change the behavior of the person who committed the offense as well as to caution the rest of the community.

Internet shaming can sometimes be effective, especially if what occurred is something that is socially unacceptable or illegal but difficult to prosecute. For example, a number of states are now posting online the names of those who haven't paid their taxes.[37] The state of Wisconsin is considering legislation to allow the posting of the names of men and women who owe back child support.[38] Internet shaming also resulted in the return of a stolen Sidekick.[39]

Despite the cases of positive effects from Internet shaming, the danger of abuse of individuals through gossip and rumors and inappropriate shaming is real. We have slander and libel laws to protect people from verbal and printed untruths, but there is nothing similar in cyberspace. The challenge we face is that there is no single entity that regulates what appears on the Internet, and most users prefer that type of freedom and openness. This issue is still evolving, and at this time there are no concrete answers.

INTERNET PATHOLOGY

Many good things have come from the communications technologies the Internet has provided. However, the Internet has also opened up some new opportunities for both criminal and psychological disturbances.

Are You Who You Say You Are?

Electronic communications isolate the user from other people with whom he or she is interacting. Although we can use video chat, communication on social networking sites and in virtual communities usually doesn't include live video. If you want others to see an image, you upload a still photo. The problem is that this makes it very easy to lie: There is no simple way to validate that you are who you say you are. For example, there is nothing to prevent an 80-year-old grandmother from uploading a picture of a trim, well-muscled, 20-something in a scanty swimsuit playing volleyball on the beach and claiming to be that person.

---

[37] http://www.usatoday.com/tech/news/2005-12-22-tax-shaming-websites_x.htm
[38] http://www.thenorthwestern.com/apps/pbcs.dll/article?AID=/20071230/OSH/312300031/1987
[39] http://consumerist.com/consumer/sidekick/sidekick-return-from-thief-after-mass-internet-shaming-182383.php

If a person restricts his or her activities to online communications and chat to benign matter—fashion, music, movies, and so on—then a false identity is usually harmless. However, when a false identity is used to get money, to harass someone, or as the basis of a relationship that will enter the physical world, that false identity can become very dangerous.[40]

One of the saddest stories of the use of a false identity on a social networking service to harass a child occurred in 2006. Megan Meier was an overweight, depressed 13-year-old with very low self-esteem who had a MySpace account. For reasons that are still unknown, she broke off a friendship with another girl who lived down the street. Shortly thereafter, she met a boy named Josh on MySpace. For a time, he flattered her, saying that she was pretty. For a lonely young teen, this was a great thing. However, suddenly Josh's communications turned ugly: He said he had shared her messages with others and that she was a bad person, fat, and a slut. The harassment continued until Megan committed suicide. Josh, however, was not a real person. He was a fake identity created by the mother of the girl who was formerly Megan's friend and a coworker. Certainly, the messages from the Josh MySpace account weren't the sole cause of Megan's suicide, but it seems that they were the proverbial "straw that broke the camel's back." Creating the false identity and sending ugly messages to Megan was not illegal, and no one has been prosecuted in the case. However, it seems unlikely that Megan would have killed herself had she not been pushed over the edge by the negative messages.[41]

News services of many kinds carry stories about child molesters who pose as young people on the Internet to lure victims to a face-to-face meeting. Molesters often flatter the potential victim, urging the child to keep the electronic contact and the eventual real-world meeting a secret. They may represent themselves as children of a similar age with similar interests, or they may tell the child that he or she is extraordinarily mature and therefore ready for a relationship with an older person. Child molesters existed before the Internet, luring children with promises of candy or requests to help find a lost puppy. However, the anonymity of the Internet has made the process much easier.[42]

Children and young people may inadvertently make it easier for a predator to target them. Postings on social networking sites that reveal personal information, even something as seemingly innocent as likes and dislikes, can provide openings that a predator can exploit. For example, if a child posts online that he likes Pokemon, a predator might pose as another child who also likes to play the game. It then becomes easy for the predator to lure the child into coming to a face-to-face card game.

---

[40] Here we are talking about representing yourself in the text of a communication as other than you really are. Providing false information when establishing an e-mail or other online account is a totally different matter and may result in jail time. See http://economictimes.indiatimes.com/Infotech/Internet_/ False_web_identity_may_land_you_in_prison/rssarticleshow/2427409.cms for one example.

[41] For complete details on this story, see http://stcharlesjournal.stltoday.com/articles/2007/11/10/news/ sj2tn20071110-1111stc_pokin_1.ii1.txt

[42] For examples of how the Internet has been used to lure children into abusive situations, see http://www.thedenverchannel.com/news/9534693/detail.html, http://news.yahoo.com/s/nm/20071026/ tc_nm/britain_chat_dc, and http://toronto.ctv.ca/servlet/an/local/CTVNews/20061124/ internet_child_luring_061124/20061124?hub=TorontoHome

Adults are also at risk from people they meet on the Internet. In late 2007, for example, a woman was murdered when she responded to a fake ad for a baby-sitting job that had been posted on Craigslist, a popular site for selling items, renting apartments, and finding job postings. She went alone to the home of the person placing the ad and apparently was murdered by the person who was in the home.[43]

This does not mean that you should never arrange face-to-face meetings with Internet friends; such meetings often work out very well. However, it does mean that you should take some sensible precautions in case the person you are meeting isn't what he or she has claimed online and you feel threatened in any way:

- Arrange the meeting in a public place, such as a restaurant or museum or heavily used park.
- Meet during the day.
- Don't go alone. Take a known friend or family member along.
- Have a cell phone handy so you can call for help if necessary.

If you are considering more than a casual friendship with someone you met online, then a background check is not inappropriate. Talk to the person's friends, coworkers, and family. It is your safety on the line, and you have a responsibility to yourself to ensure that the person is how he or she appeared online.

The Internet and Addictions

We can define an addiction as anything that a person cannot do without, whether it be physical or psychological. It is something that a person has a compulsion to do and that is very difficult (if not impossible) to stop. An addiction also interferes with normal daily functioning. Certainly, the Internet doesn't create physical addictions like tobacco and some drugs, but can someone become psychologically addicted to using the Internet? The answer depends on which research you are reading.[44]

The idea of Internet addiction began as a hoax.[45] Ivan Goldberg, a psychiatrist, proposed the disorder as a joke in 1995. However, since that time some serious research has been conducted on the possibility that people can become addicted to use of the Internet (shopping, playing games, chatting with other people, and so on). Internet "overuse" does show many of the characteristics of other psychological addictions (for example, compulsive gambling or shopping): performance of the undesirable behavior to such a degree that it interferes with normal personal and job functioning. Nonetheless, professional associations, such as the American Medical Association, have yet to label technology overuse as addiction.[46]

Despite the lack of formal recognition of Internet addiction as a disorder, some professionals encounter those who have profound problems with removing themselves from the computer. For example, some Chinese believe that as many as 15% of Chinese teenagers suffer from some form of the disorder.[47] Those

---

[43] For details see http://www.foxnews.com/story/0,2933,306317,00.html

[44] For a fun test to see if you're Internet-addicted, go to http://www.netaddiction.com/resources/internet_addiction_test.htm

[45] See http://www.nurseweek.com/features/97-8/iadct.html for details.

[46] See http://psychcentral.com/news/2007/06/26/video-games-no-addiction-for-now/

[47] See https://www.mywire.com/Auth.do?extId=10022&uri=/archive/forbes/2005/0509/054.html and http://observer.guardian.co.uk/international/story/0,6903,1646663,00.html

affected have dropped out of school and become violent when forced to give up Internet use. At the very least, they show decreased school and job performance; their interactions with family and friends suffer significantly.

One question we need to answer is whether Internet addiction is a significant danger to our world. Consider that some people have what we could call "addictive personalities" (or, perhaps, "addictive personality disorders"). These are people who are likely to become addicted to *something* that will isolate them from the problems of everyday life. Perhaps they would drink too much, gamble too much, or take addictive drugs. Instead, they become compelled to spend time using the Internet. At this time, there is no evidence that Internet addiction has significantly increased the number of addicted individuals. Instead, people become addicted to the Internet instead of becoming addicted to something else.

On the other hand, Internet use is cheaper than any other addiction—it requires only a computer and a connection to the network—and therefore is available to more people than compulsive gambling or drug abuse. It also does not compel the same type of behavior as addictions that cost more because it usually doesn't represent a significant drain on personal or family finances. Internet overusers do not need to resort to theft to support their habits. Heavy computer users tend to be young and for the most part don't have high incomes or the means to acquire large amounts of money. If they are going to become addicted to something, the Internet and other technology use is something that is easy to abuse.

Dangers to Kids

Earlier in this section we talked about the use of the Internet by sexual predators to lure children into face-to-face meetings. Although this is probably the greatest danger facing children who use the Internet, there are others. In particular, children may be exposed to Web sites with content that is either aimed at adults or factually inaccurate. They may also be subject to privacy violations by Web sites that attempt to collect personal information about underage users.

One of the things we enjoy about the Internet is that it isn't censored, that it is open to just about anything anyone wants to post. (The only thing that seems to be universally banned on the Internet is child pornography.) Unfortunately, that also means that it contains content that most of us would consider inappropriate for children. This includes hate sites, adult pornography, and instructions on building bombs and how to commit murder. The freedom of the Internet therefore requires that children be protected until they are able to make their own decisions as to which content is appropriate.[48]

Children are often unable to discriminate between propaganda and legitimate sources of information. Given that so many young people turn to online sources for research rather than using more traditional print sources, they need help to evaluate the accuracy and biases of what they are reading. Parents have

---

[48] As a starting point for research on how to protect children when they are using the Internet, see http://www.cbsnews.com/stories/2003/05/07/earlyshow/living/parenting/main552841.shtml. In particular, follow the "Related" links to interactive content.

always been responsible for monitoring what their children see, hear, and read. The Internet has not added anything new in that respect. However, it does mean that parents need to be somewhat computer savvy to be aware of what their children are doing online. At this time, there are parents of young children who are technology resistant and who may be placing their children at risk when they are online. However, as we have discussed in this book, as the current population of children and teens matures into adults, there will be far fewer people who haven't grown up with computers, somewhat alleviating the problem. What even the most vigilant parents cannot eliminate is the inappropriate and inaccurate content; therefore they must prevent their children from reaching inappropriate content and teach their children to evaluate information to determine, at least in some measure, its accuracy.

Computer Crime: Old Crimes With New Tools

To round out our discussion of social problems caused by misuse of stored and transmitted data, we should take a quick look at plain old computer crime: Crimes that are committed using computers. The Internet has made it easier to commit a number of crimes, although if we look carefully, none of the crimes is really new.

The Internet has made it easier to engage in theft of information, whether for industrial espionage or spying. An information thief no longer has to visit the offices where the information he wants to steal is stored; the theft can be done remotely over a computer network. In many ways, this reduces the risk for the thief or spy—there are no alarm systems or security guards to worry about—but also requires a new type of expertise. Just knowing how to get in quietly through a window isn't enough! The electronic information thief must either be very knowledgeable, have an account name and password for the target computer, or have a colleague who is very knowledgeable.

The "good" computer thief can get in and out of a target system without leaving a trace behind. Some computer crime is detected and reported publicly. For example, in 2001 the computer system of a bank in Waco, Texas was penetrated by Russian hackers.[49,50] Chinese hackers broke into a railroad's computer system.[51] And personal information was gleaned by hackers from ChoicePoint, an organization that collects background information on many U.S. citizens.[52] However, the really important break-ins are those we never hear about, those performed by thieves so talented that no on ever knows a crime has occurred.

The Internet also makes it easy to commit vandalism. Hackers gain access to a Web site, usually one supporting political or ethical positions with which the hackers don't agree, and then modify the Web site to either reflect their own positions or ridicule the position of the Web site operators. In broad terms, defacing a

---

[49] http://www.landfield.com/isn/mail-archive/2001/Jun/0010.html

[50] The term "hacker" is a funny one. If we are being rigorous, a hacker is someone who does something very clever with electronics; in some cases, calling someone a good hacker is a compliment and saying that a piece of software or hardware is "a good hack" means that the item is very, very well done. However, the term has taken a pejorative meaning when associated with illegal access to computer systems. True hackers prefer that those who break into systems illegally be called "crackers."

[51] http://lists.jammed.com/ISN/1999/01/0111.html

[52] http://www.findbase.com/headlines/comp-cp-break-in-imperils-ca-sfgate-2.16.-05.htm

Web site electronically is no different from spray-painting something on a wall or defacing a poster. For example, in 2004 a group of hackers accessed and modified the Al-Qaeda Web site.[53] They redirected traffic intended for the legitimate site to one of their own, where they displayed an antiterrorist threat.

THE LEGAL LANDSCAPE

For a long time it wasn't clear to lawmakers how they should deal with crime related to electronics. Many thought that existing laws could be adapted to technological crime. For example, if someone broke into a computer, then he or she would be prosecuted under the statutes governing breaking and entering. As the use of electronic communications and electronic data storage grew, it became clear in the United States that new laws were needed to deal with electronic crimes and the privacy of electronic information.

Crimes Against Electronic Information

Crimes against electronic information are defined in Title 18/Part I of the U.S. legal code.[54] Chapter 119 of the law deals with electronic communications; Chapter 121 covers stored electronic communications and data.

The purpose of Chapter 119 is to protect e-mail and other types of transmissions. It prohibits the "manufacture, distribution, possession and advertising of wire, oral, or electronic communication interception devices." The statute also lays out the requirements for legal interception of such transmissions—what we might call "wire tapping"—and the penalties for illegal interception.

The provisions of this law can be hard to enforce. For example, many network administrators use a "packet sniffer," a device that captures packets as they travel over a network. This can be a legitimate activity for monitoring the health of a network. However, a packet sniffer can also be used to intercept network traffic on networks the operator isn't authorized to sniff and, in the hands of a knowledgeable user, reveal the contents of a message. Therefore, is a packet sniffer a prohibited device? It's not—and the U.S. law makes special provision for the use of otherwise prohibited activities and devices for the providers of an electronic communications service—but like many network maintenance devices (both hardware and software) a packet sniffer has both legitimate and illegal uses.[55]

Chapter 121 does for stored data what 119 does for data in transit: It lays out which actions are illegal and the penalties for those actions when dealing with stored electronic communications. In addition, Chapter 121 covers telephone records and video tape rental and sale records. Notice that these laws cover only

---

[53] http://www.techweb.com/wire/security/48800108

[54] See http://www4.1aw.cornell.edu/uscode/html/uscode18/usc_sup_01_18_10_I.html for a readable version of the laws. Title 18 contains laws governing all types of federal crimes, from Aircraft through War Crimes.

[55] The other thing to keep in mind about packet sniffing is that even if someone were to capture packets traveling on the Internet, most messages are split into multiple packets, all of which may not travel the same path from source to destination. In addition, there is so much traffic on the Internet that capturing all the packets from an individual message could be extremely difficult. These factors alone make messages in transit through the Internet relatively secure; data are far more vulnerable when they are stored on the hard disk of a computer attached to a network.

the stored form of electronic communications and the telephone and video tape rental transactions. They do not deal with data that were generated in any other way, such as orders taken over the telephone and stored in a database.

What we typically think of as "hacking"—unauthorized access to any computing resource—is governed in the United States by the Computer Fraud and Abuse Act.[56] In was originally enacted in 1984 and amended as late as 1996. Found in Title 18/Part I, Chapter 47 (laws against fraud of all types), section 1030 deals with the following abuses:

- Intentionally accessing a computer to which one has no authorization to access
- Exceeding one's authorized access to a computer system
- Using such illegally obtained information to commit fraud or extortion of any kind (This provision certainly covers phishing and other such activities.)
- Causing any kind of damage (including physical and financial) through unauthorized access (Malware that causes losses of any kind can be prosecuted under this provision.)

Although we typically think of the U.S. Secret Service as the agency that protects U.S. federal officials, it is also the branch of the federal government given investigative authority in cases of computer crime. The FBI becomes involved when the computer crime involves espionage, foreign intelligence, and national security.

In addition to the federal statutes, each of the U.S. states has its own laws governing computer hacking.[57] For example, Nevada has enacted statutes that equate computer hacking to fraud. The same statute covers identity theft.[58]

The United States is not alone in enacting laws to cover computer crime. Great Britain, for example, passed the Computer Misuse Act 1990, which is substantially similar to laws in the United States.[59] More recently, the European Union adopted the "Framework Decision of Attacks Against Information Systems." The framework identifies three criminal offenses:

- Illegal access to information systems
- Illegal system interference (the intentional serious hindering or interruption of the functioning of an information system by inputting, transmitting, damaging, deleting, deteriorating, altering, suppressing or rendering inaccessible computer data)
- Illegal data interference

Although the framework was adopted in 2005, members of the European Union were given until mid-March 2007 to implement its provisions.[60] Because the European Union's framework doesn't have the force of law, the framework specifies that each member country must enact its own laws to cover punishments for crimes committed under the framework.

---

[56] For the text of the law, see http://www4.1aw.cornell.edu/uscode/html/uscode18/usc_sec_18_00001030—000-.html
[57] For links to the state laws, see http://www.ncsl.org/programs/lis/CIP/hacklaw.htm
[58] See http://www.leg.state.nv.us/NRS/NRS-205.html#NRS205Sec473
[59] For text of the act, see http://www.opsi.gov.uk/acts/acts1990/Ukpga_19900018_en_1.htm
[60] The full text of the framework can be found in English at http://europa.eu/scadplus/leg/en/lvb/l33193.htm

The bottom line is that technology has opened new avenues of crime to which government bodies have responded by enacting the laws they deem necessary to handle such crime. Although the tools of the crimes are new—electronic devices and software—the intent of the crimes is not. People are still attempting to steal trade secrets by accessing research and development information, prevent others from expressing their points of view by defacing Web sites, engage in vandalism, and so on. The methods may be different, but the intent—and often the results—are the same.

Privacy Laws

Privacy laws in the United States include those that not only restrict what information stored by the federal government can be released, but the information to which an individual has access. Other laws govern what retailers may and may not do with the data you give them when making a purchase as well as the accessibility of educational records.

The U.S. government has no direct authority over private educational institutions. However, what it does wield is considerable power over federal funds that are allocated to schools. Therefore the U.S. laws that govern the privacy of educational records deny federal funds to those schools that don't meet the requirements of the law. The primary law governing the privacy of school records is the Family Educational Rights and Privacy Act (FERPA).[61] Originally enacted in 1974, FERPA has been amended nine times. It states that parents have a right to view the records of children younger than 18 years of age and that the records of older students (those 18 and over) cannot be released to anyone but the student without the written permission of the student. Schools therefore have the responsibility of ensuring that student records are not disclosed to unauthorized people, thus the need for secure information systems that store student information.

The U.S. federal legal code also includes the Health Insurance Portability and Accountability Act (HIPA).[62] Like FERPA, its intent is to safeguard the privacy of medical records. It restricts the release of medical records to the patient alone (or the parent/guardian in the case of those under 18) or to those the patient has authorized in writing to receive the records. It also requires the standardization of the formats of patient records so they can be transferred easily among insurance companies and the use of unique identifiers for patients. (The social security number may not be used.) Finally, the law requires that security measures be in place to protect the privacy of medical records.[63]

---

[61] See http://www.ed.gov/policy/gen/guid/fpco/ferpa/leg-history.html

[62] See http://www.hipaadvisory.com/REGS/HIPAAprimer.htm

[63] For some medical professionals, HIPAA is sometimes a major problem. Consider the following not-so-hypothetical situation related to me by a doctor: The doctor had a patient who was mentally incompetent to make his own medical decisions. The doctor wanted to go to court to have a medical guardian appointed for the patient. To do that, the court needed access to the patient's medical records, but HIPAA wouldn't allow the records to be released to the court without the patient's informed consent. But the patient wasn't able to give informed consent, which is why the court was involved in the first place. Catch-22! Either the doctor breaks the law and releases the records to the court without the written consent of the patient or the doctor may be unable to administer needed treatment to the patient.

An area that has been given special attention is the privacy of children. The Children's Online Privacy Protection Act of 1998 governs that data can be requested from children (those under 13) and which of those data can be stored by the site operator.[64] It applies to Web sites, "pen pal services," e-mail, message boards, and chat rooms. In general, the law aims to restrict the soliciting and disclosure of any information that can be used to identify a child—beyond information required for interacting with the Web site—without the approval of a parent or guardian. Covered information includes first and last name, any part of a home address, e-mail address, telephone number, social security number, or any combination of the preceding. If covered information is necessary for interaction with the Web site—for example, for registering a user—the Web site must collect only the minimally required amount of information, ensure the securing of that information, and not disclose it unless required to do so by law.

To comply with the law, many Web sites that collect personal information now require users to assert that they are at least 13 years old. The problem with this, of course, is that although such an assertion complies with the letter of the law, it does nothing to prevent a younger child from stating that he or she is 13. In contrast, Web sites that cater to children either don't require children to register with the site at all or collect minimal information such as a first name and birth date. The child is then assigned an arbitrary user name and password.[65]

BIG BROTHER: TRACKING OUR POSSESSIONS AND US

In 1932, Aldous Huxley wrote *Brave New World*, a book that has come to symbolize the horror of a totalitarian state where no one has any privacy. Spying on citizens is conducted by a technological intelligence known as "Big Brother." That term now represents anything that tracks the movements and behaviors of human beings. For the most part, it is considered to be very negative.

In North America and Europe we consider ourselves generally free from constant observation. However, although we may not think of it regularly, many of our daily activities are not anonymous and allow our movements to be tracked. Consider the following:

- Every time you do anything at an ATM you place yourself in time and space.
- Every time you use a credit card you place yourself in time and space.
- If you use E-Z Pass, the automated toll paying technology that is widespread in the northeastern United States, each time you pass through a toll booth you place yourself in time and space.[66]
- Using a cell phone also can place you in space and time. (Your exact location can be determined by triangulating between cell towers.)

---

[64] See http://www.ftc.gov/ogc/coppa1.htm for the full text of the law.
[65] As an example, sign up for an account at www.cheetos.com or http://www.cartoonnetwork.com/tv_shows/fosters/index.html
[66] See http://www.ezpass.com/

It's almost impossible to live in our society and not leave digital footprints behind somewhere. Until recently, most of the trails we've left behind have come from financial transactions. A new technology has entered the arena, however: the radiofrequency identification (RFID) tag.

An RFID tag is a small computer chip that broadcasts an identification code that can be picked up by a receiver in close proximity. The code can then be used to retrieve information about the tagged item from a database. Initially, there were two planned uses for the chips: inventory control and the identification of animals. Products that carry RFID tags can be inventoried without physically moving the items from a storage location.[67] In fact, RFID tags could lead to automatic grocery store checkout: Put the tagged goods in your cart, insert the loaded cart in an RFID scanner, and wait for the scanner to identify your purchases.

We regularly embed chips in our pets so they can be returned when lost; farmers embed the tags to track farm animals. For example, RFID tags can make it easier to track the movements of a cow infected with mad cow disease. The U.S. Department of Agriculture has mandated the use of RFID tags for all livestock, beginning in 2009.[68] In most cases, people have very few objections to these uses of technology.

The controversy arises when we consider the possibility of embedding these chips in people. For those with medical problems who might otherwise wear a MedicAlert bracelet, RFID chips can give medical professionals almost instant access to medical records.[69] Consideration has been given to using the chips as well in patients with diminished mental capacity who are prone to wander. The concept has also been extended to corporations whose security needs are such that they must know the location of every employee at all times.

The idea of implanting RFID chips in people conjures up visions of Big Brother for many people. However, proponents of the tags cite the following positive uses:

- Medical records are available even if a person is unconscious.
- Financial transactions can be completed without carrying credit cards. The chip's ID number links to bank accounts from which money can be deducted. The most widely known financial use today is a resort in Spain that tags its guests so they don't have to carry ID cards with them and can pay for meals and other items by passing their wrists across a scanner.[70]
- The movements of convicted sex offenders who have been released from prison can be tracked. It would be simple, for example, to determine that a pedophile was too close to a school.
- People can voluntarily participate in groups that allow tracking of members' movements so that all members of the group can stay in contact.[71]
- If all children were tagged, missing children could be identified, regardless of how long they had been missing.[72]

---

[67] For a sample product and associated services, see http://www.activewaveinc.com/applications_inventory_control.php
[68] For more information, see http://technocrat.net/d/2006/4/7/2165
[69] You can see the story of someone who chose to have the chip implanted at http://www.networkworld.com/news/2005/040405widernetchip.html
[70] http://www.bajabeach.es/
[71] http://www.networkworld.com/news/2005/040405widernetchip.html describes a pilot project about such a use of RFID.
[72] For some additional positive uses of RFID technology, see http://www.komotv.com/news/4594356.html

However, there are a number of objections to using the chips in people:

- People have a right to privacy, and the RFID chips can be misused to violate an individual's right to private actions. Anecdotal evidence suggests that most people consider the privacy issue to be the overriding argument against implanting tags in people.[73]
- The chips can be read by any reader that can access a chip's frequency. There is currently no way to keep hackers from reading the ID codes from the chips and then hacking into the associated database to retrieve someone's stored data.[74]
- There is no way to be certain at this time whether the chips are safe to implant in a human body. There are questions about the long term safety of the materials out of which the chips are made as well as the effects of the radio waves emitted by the chips.

Given the short time that RFID tagging has been available for humans, it is still unclear whether the use of the technology will become widespread. The debate is likely to continue for a number of years.

---

[73] http://www.news.com/2010-1039-5332478.html
[74] For a discussion of the privacy issues surrounding RFID chips, see http://www.fcw.com/print/12_18/news/94595-1.html

WHERE WE'VE BEEN

In this chapter, we have considered the effects of technology on human interactions and behavior. Electronic communications, beginning with the telegraph, have brought people and commerce closer together. Electronic communications are well accepted by younger generations and have replaced older forms of communication, such as letter writing. Although many older people are uncomfortable with communications vehicles such as e-mail, text messaging, and blogs, such activities help people keep in touch. Because writing e-mail is easier for most people than writing a letter, e-mail tends to promote interpersonal relationships rather than hinder them.

The Internet has made possible new forms of human interaction, including social networking and virtual communities. These Web-based applications let people form relationships with others throughout the world.

Technology has also made it easier for people to gain access to information that we would prefer to keep private, such as financial data and data that identifies us. The biggest threat from this loss of privacy is identity theft, through which someone illegally uses another person's credentials to access money and gain credit.

Electronic communications and data storage have also created their own pathologies. There are new types of crime associated with unauthorized access to computer systems. Countries around the world have enacted legal statutes to define and punish crimes against electronic communications and stored data.

The anonymity of the Internet makes it possible for people to represent themselves online as something they are not. The biggest danger from this is to children, who can be lured into face-to-face meetings with sexual predators. To help protect children, the United States has enacted a law that prevents the soliciting and storage of information that can identify a specific child and where that child is located.

Some individuals overuse computing resources such as the Internet. Their behavior suggests an addiction to computer use.

THINGS TO THINK ABOUT

1. An e-mail is not the same as a letter sent by first class mail. In particular, it lacks the style of a letter. Assuming that this change is acceptable to most people, do you think we have lost something culturally by moving to a more abrupt style? Why or why not?

2. Once you pay fees to your ISP, much of what you do on the Internet is free. Instant messaging is free, for example; Google searches are free. These activities are supported by advertising, just as are many television and radio stations. Is this a good way to fund online activities? Why or why not? If you aren't in favor of advertising, what would you suggest as an alternative?

3. Consider the problem of identity theft. What are some things that each individual can do to prevent becoming a victim?

4. Suppose that you meet someone over the Internet who seems like a decent person that you would like to know better. Before arranging to meet this person, what can you do to verify that the person is who he or she claims to be?

5. Research the problem of Internet addiction. Assuming that it is indeed a real disorder, what are its symptoms? How is it treated? In what ways is it similar to other addictions? In what ways is it different? Do you know anyone that you would consider to be addicted to computer or technology use? If so, describe the person's behaviors that lead you to think so.

6. You read in this chapter that many government entities have enacted laws to cover computer crime. Take a look at the text of several computer crime laws and at laws against theft of physical property and breaking into places where a person isn't authorized to be. How are the computer laws similar to the physical laws? How are they different? Do the results of your analysis suggest that the specific laws against computer crime were warranted, or could computer crime have been handled by the older laws? Why?

7. In this chapter, we discussed some ways in which computer crimes are similar to crimes in the real world but with different tools. In what ways are computer crimes different from physical crimes? Does this make the crimes any less illegal or ethically wrong? Why?

8. The Internet has made it possible for groups that espouse terrorism and hate, such as Al-Qaeda, to have a voice equal to that of anyone else in the world. Should groups such as this be allowed on the Internet? Why? If so, is it ethically correct to hack into those sites and modify them to dilute the message of hate? Why? If not, who should have the responsibility of determining what should be posted and what should be forbidden?

9. What would it take to live totally "off the grid" in a Western society? What kind of lifestyle would you need to prevent anyone from knowing anything about where you were and what you did?

10. In this chapter, the section of rumors, gossip, and shaming ends with the statement that currently there are no concrete solutions to the problem of the dispersal of information about the behavior of individuals. What could or should the Internet community do to prevent the lives of people from being ruined by disproportionate spread of rumors and so on? Should governments become involved? If so, how? If not, why not?

WHERE TO GO FROM HERE

Communications

Green, Lelia R. *Communication, Technology and Society*. Thousand Oaks, CA: Sage Publications, 2002.

"The Impact on Society." *http://www.connected-earth.com/Galleries/Transformingsociety/Theimpactonsociety/index.htm*

Kaupplla, Amanda. "Advances in Technology Affect Interpersonal Communication." *http://media.www.theguardianonline.com/media/storage/paper373/news/2007/02/07/News/Advances.In.Technology.Affect.Interpersonal.Communication-2704052.shtml*

Thomas, Matthew. "The Impacts of Technology on Communication—Mapping the Limits of Online Discussion Forums." *http://online.adelaide.edu.au/LearnIT.nsf/URLs/technology_and_communication*

Social Networking and Virtual Communities

Davis, Donald Carrington. "MySpace Isn't Your Space: Expanding the Fair Credit Reporting Act to Ensure Accountability and Fairness in Employer Searches of Online Social Networking Services." 16 *Kansas Journal of Law and Public Policy* 237, 2007.

"Diplomacy Island in Second Life." *http://www.diplomacy.edu/projects/vd/sl.asp*

Dodero, Camille. "Lost in MySpace: Log on, tune in, and hook up with 22 million people online." *The Boston Phoenix,* July 22–28, 2005.

"Facebook Opens Profiles to Public." *http://news.bbc.co.uk/2/hi/technology/6980454.stm*

Jones, Harvey, and Jose Hiram Soltren. "Facebook: Threats to Privacy." *http://www.swiss.ai.mit.edu/6095/student-papers/fall05-papers/facebook.pdf*

Kirkpatrick, David. "Facebook's Plan to Hook up the World." *http://money.cnn.com/2007/05/24/technology/facebook.fortune/*

Lash, Devon. "Site Used to Aid Investigations." *http://www.collegian.psu.edu/archive/2005/11/11–10–05tdc/11–10–05dnews-09.asp*

Mitchell, Dan. "What's Online: The Story Behind MySpace." New York Times Online. *http://query.nytimes.com/gst/fullpage.html?res=9E01E3DE1331F935A2575AC0A9609C8B63*

"A Second Look at School Life." *http://education.guardian.co.uk/elearning/story/0,,2051195,00.html*

"Top 3 Act-and-Look-Alikes: Second-Life Avatars and their Real Life Users." *http://weburbanist.com/2007/06/17/top-3-look-alike-avatars-and-people-from-second-life-to-real-life/*

Vance, Ashlee. "Google Pays $900m to Monetize Children via MySpace." *http://www.theregister.co.uk/2006/08/07/google_wins_myspace/*

Interpersonal and Family Relationships

Barnes, Susan B. *Online Connection: Internet Interpersonal Relationships*. Creeskill, New Jersey: Hampton Press, 2001.

DeVito, Joseph. *The Interpersonal Communication Book (11th ed.)*. Upper Saddle River, New Jersey: Allyn & Bacon, 2006.

"Family Relationship Article." *http://tntn.essortment.com/familyrelations_rcgb.htm*

Holder, Daniel S. "Ethnographic Study of the Effects of Facebook.com on Interpersonal Relationships." *http://laurenashleigh.blogspot.com/2005/07/how-modern-day -technology-is-ruining_08.html*

Hutchby, Ian. *Children Technology and Culture: The Impacts of Technologies in Children's Everyday Lives*. New York: Routledge Falmer, 2001.

McQuillen, Jeffrey S. "The Influence of Technology on the Initiation of Interpersonal Relationships." *http://findarticles.com/p/articles/mi_qa3673/is_200304/ai_n9232834*

Sullivan, Bob. "Deaf Turn to Dating Online." *http://www.msnbc.msn.com/id/3078714/*

Internet Pathology

Briggs, Rudolph G. "Psychosocial Parameters of Internet Addiction." *http://library.albany.edu/briggs/addiction.html*

Grohol, John M. "Internet Addiction Guide." *http://psychcentral.com/netaddiction/*

Illinois Institute for Addiction Recovery. "Internet Addiction." *http://www.addictionrecov.org/internet.htm*

Stadler, Jen. "Internet Safety News: Online Child Molesters." *http://www.netsmartz.org/news/onlinemolesters.htm*

"Tel Aviv University Redefines 'Internet Addiction' and Sets New Standards for Its Treatment." *http://www.tauac.org/site/News2?page=NewsArticle&id=5739*

Law, Privacy, Security, and Computer Crime

Britz, Marjie T. *Computer Forensics and Cyber Crime: An Introduction (2nd ed.)*. New York: Prentice Hall, 2008.

Computer Crime and Intellectual Property Section, U.S. Department of Justice. *http://www.cybercrime.gov/*

Computer Crime Research Center. *http://www.crime-research.org/*

Cyber Investigations, Federal Bureau of Investigation. *http://www.fbi.gov/cyberinvest/cyberhome.htm*

Cybercrime Law. *http://www.cybercrimelaw.org/index.cfm*

"Cyberspace Law." *http://www.findlaw.com/01topics/10cyberspace/computercrimes/index.html*

"Fighting Back Against Identity Theft." *http://www.ftc.gov/bcp/edu/microsites/idtheft/*

Hallam-Baker, Phillip. *The dotCrime Manifesto: How to Stop Internet Crime*. Indianapolis, Indiana: Addison-Wesley Professional, 2007.

Solove, Daniel J. *The Future of Reputation: Gossip, Rumor, and Privacy on the Internet*. Princeton, New Jersey: Yale University Press, 2007.

Standler, Ronald B. "Computer Crime." *http://www.rbs2.com/ccrime.htm#anchor666666*

Wall, David S. *Cybercrime*. Cambridge, England: Polity, 2007.

Government, Politics, and War 8

WHERE WE'RE GOING

In this chapter we will

- Explore how governments handled information and documents before the introduction of information technology.
- Consider some of the major ways in which governments use technology for their day-to-day functioning.
- Look at government surveillance of e-mail traffic.
- Discuss the influence of television on politics.
- Explore the issues surrounding the use of modeling software and exit polls to predict the results of elections.
- Discuss how technology has changed political campaigns.
- Look at the technology used in some major 20th- and 21st-century wars.
- Consider military technologies that have had civilian benefits.

INTRODUCTION

For good or for ill, humans have had some form of government, politics, and war since the beginning of civilization. We have been able to manage all of them without technology, with varying degrees of success. However, technology has certainly changed how we govern ourselves, choose our leaders, and make war.

GOVERNMENTS AND TECHNOLOGY

Given the vast amount of information handled by world governments, governments and information technology would seem like an ideal pairing. Some countries have adopted significant amounts of technology in their national and local governments, whereas others (like the United States) face some barriers to the adoption of technology throughout the government.

Government Functioning Before Information Technology

Before the 1960s, when computer analysis of government data became feasible, governments ran on paper—lots of paper. The initial U.S. Constitution was transcribed by hand by a clerk named Jacob Shallus,[1] but about 20 copies of the four-page document (**Figure 8.1**) were made shortly thereafter. From that moment on, the volume of paper produced by official functions of the U.S government has grown exponentially. Recent budget documents regularly appear in multiple volumes that run to thousands of pages.[2] The documents are available online as PDF files, but members of Congress still receive paper copies.[3] The same is true for proposed legislation: Although electronic versions are available, most government offices rely on paper for day-to-day work.

Initially, governments had to make document copies by hand. The development of plain-paper photocopying (**Figure 8.2**) in the late 1950s changed all that: Suddenly, it was possible to make multiple copies of documents easily and quickly. The amount of paper used by governments exploded.

Government Uses of Technology

Governments were among the first users of computing technology. For example, the first commercial computer—UNIVAC—was used by the U.S. government to process data from the 1960 census. At least in the United States, however, the introduction of technology into government departments has been relatively piecemeal. Each government department has developed its own technology. The Internal Revenue Service (IRS) was an early adopter, photographing federal tax returns along with storing the data electronically. Today, it allows field agents to connect to home systems through a virtual private network, or VPN.[4] IRS technology has assisted in identifying those channeling funds for terrorists by examining the movement of money.[5] Similar systems identify taxpayers to audit by comparing deductions with income.[6] The agency is also moving away from the filing of tax returns on paper by encouraging people to file electronically.[7]

---

[1] The identity of the person who physically produced the first copy of the U.S. constitution wasn't known until 1937. For more constitutional trivia, see http://www.archives.gov/national-archives-experience/charters/constitution_q_and_a.html

[2] The United States has enacted a "Paperwork Reduction Act," but the law isn't exactly what its name implies. It has nothing to do with reducing the amount of paper used by the government; it deals with reducing the amount of paperwork that people and organizations need to complete to supply information required by the government. For more information, see http://www.archives.gov/federal-register/laws/paperwork-reduction/3501.html

[3] You can find PDFs of the current U.S. budget at http://www.whitehouse.gov/omb/

[4] A VPN provides security for private communications over the Internet. To read more about IRS use of this technology, see http://findarticles.com/p/articles/mi_m0EIN/is_2001_Jan_22/ai_69373476

[5] For details, see http://www.gcn.com/print/25_6/40153-1.html

[6] A person who reports a gross income of $30,000 and a deduction of $15,000 for mortgage interest is probably hiding something, for example.

[7] The agency also seems to have a security problem with unauthorized wireless access points installed by employees. For details see http://findarticles.com/p/articles/mi_kmusa/is_200704/ai_n19003640

Figure 8.1
The first page
of the original
U.S. Constitution.

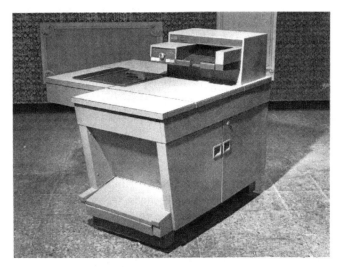

Figure 8.2
The Xerox 914,
the first automatic
plain-paper copier
introduced in 1959.

Some governments have embraced technology as policy, such as Estonia (discussed in Chapter 7). Australia has been moving toward what it calls "e-Government" since 2002. The intent is to make government more responsive to the population.[8] In addition, Australia is moving to integrate the many databases that store government information. Other countries, such as Canada, also have a commitment to increase the amount of technology in use in government and are moving slowly toward e-government.[9]

There is, however, a barrier to the adoption of the latest technology in the United States: the legislative process. For example, when the IRS asked for new computing systems in the 1990s, they had to write specifications and obtain bids from vendors. Then, the information went to Congress, where it was debated as part of a spending bill and finally approved. It took approximately three years from the time the IRS obtained its bids until the money was actually available. By that time, the technology in the system plan was outdated.[10]

Government Record Keeping Governments have always kept records about their citizens. Some were as simple as tax rolls; others were more sinister, such as data gathered from clandestine surveillance. Much of these data were kept in manila file folders, requiring "librarians" to help government workers retrieve data they wanted. In the years before World War II, the Nazis were able to identify Jews using existing paper records, which were stored in shoe boxes.

The consolidation of databases in electronic form has been both negative and positive:

- Governments can consolidate social program benefits, in the process detecting fraud to save taxpayers money.
- Tax collection bureaus can cross-match sources of income with tax returns to identify those who fail to report income. Whether this is a good thing depends on your point of view. It can ensure that everyone pays what is owed, reducing the tax burden on everyone and ensuring financial stability of the country. On the other hand, if you happen to be someone who wants to hide income so you pay less taxes, the ability to match income to tax returns is a definite problem.
- Governments can consolidate intelligence data to better identify terrorists and other individuals who are working against the country. Law enforcement agencies can share data to help solve crimes and locate wanted individuals.
- Governments can accumulate data that can be used to restrict personal freedom. Such data might include records of antigovernment writings, appearances at political gatherings, people with whom individuals associate, library borrowing records, transcripts of telephone conversations, e-mail messages, and so on.

---

[8]For details, see http://www.apsc.gov.au/mac/technology.htm
[9]For information on the Canadian systematic approach to applying technology to national government, see http://www.tbs-sct.gc.ca/fap-paf/documents/iteration/iteration04_e.asp
[10]The IRS's project did go ahead, but it was not successful. The problem actually had nothing to do with the legislative time lag, however. Poor management resulted in significant delays in the delivery of completed systems. See http://findarticles.com/p/articles/mi_m5072/is_9_25/ai_98945785 for more information.

Technology plays a major part in the accumulation of the data stored in those databases. Through the 1980s, most U.S. government databases were stored on reel-to-reel tape. There was no easy way to consolidate or cross-reference information between them. However, the development of database management systems (DBMSs) and the reduction in the cost of disk storage made it possible to have multiple databases that were easy to search online at the same time.

Mail Surveillance We put a lot of things in our e-mail messages, usually without thinking about what might happen if someone other than the recipient read it. Although it is unlikely that a hacker can intercept and read the message while it is in transit, the same is not true for government agencies. For example, U.S. government software (originally named Carnivor and now called DCS-1000) regularly intercepts e-mail coming to and from international locations. The volume of e-mail is too high for the software to examine every message, so it scans messages looking for keywords such as "terror" or "bomb."

There is minimal social resistance in the United States to the surveillance of international e-mails, but domestic surveillance is an entirely different matter. For the most part, the U.S. courts have viewed intercepting domestic e-mails as similar to wiretapping phones: A warrant is required to do so. In contrast, some court cases have required organizations to produce copies of e-mails as evidence. For example, during the period when it was fighting antitrust suits, Microsoft was ordered to produce a number of e-mails that purportedly contained discussions of issues relevant to the court cases. Microsoft was unable or unwilling to do so, claiming that the e-mails had been destroyed or that the volume of e-mails received at Microsoft made it too hard to search for specific messages.[11] The wrangling over the need to produce old e-mail messages in legal matters has meant that organizations must now be much more diligent in keeping archives of e-mails, rather than deleting them regularly. There is always a "chance" that an e-mail message could be subpoenaed as evidence in a court case. It also means that people should be careful what they write when using their employers' e-mail systems: Anything put down electronically is as liable to be considered evidence as anything stored on paper.

Governments as Technology Providers

As well as using technology, some governments have gone into the business of providing Internet access to their citizens. Some communities are providing wireless broadband, whereas others are supporting fiber optic networks. **Table 8.1** contains the status of municipal broadband access in the 50 United States as of January 2008. Notice that most of the existing systems are in smaller communities, although systems are planned or under consideration in larger cities. There also does not seem to be a correlation between the size, population, or wealth and the number of communities with municipal broadband access.

Existing systems are in plain type; forthcoming systems or systems being considered are in italics; empty spaces indicate no existing or planned systems.

---

[11] See http://www.technologyevangelist.com/2007/02/microsoft_dirty_tric_1.html

Table 8.1 Status of Municipal Broadband Access in the United States

State	Communities/Regions With Wireless Systems	Communities/Regions With Fiber Optic Systems
Alabama	*Childerburg, Cullman, Huntsville, Madison*	Sylacauga
Alaska*	*Juneau, Kodiak*	
Arizona	Rio Rico, Scottsdale, Tempe, Tucson; *Flagstaff, Phoenix, Kingman, Mesa, Sahuarita, Yuma*	
Arkansas		
California	Burbank, Cerritos, Culver City, Cupertino, Encinitas, Fullerton, Hermosa Beach, Livermore, Long Beach, Lampoc, Los Angeles, Milpitas, Mountain View, Pleasanton, Riverside, San Diego County, San Mateo, Temecula; *Alameda, Anaheim, Azusa, Belmont, Burlingame, Covina, East Palo Alto, Folsom, Foster City, Fresno, Gustine, Hillsborough, Irvine, Los Banos, Los Gatos, Modesto, Oakland, Pacifica, Pasadena, Pleasant Hill, Ripon, Sacramento, San Diego Indian tribal village, San Francisco, San Jose, Santa Clara, Simi Valley, Sunnyvale, West Hollywood*	Loma Linda; *Fontana, Lompoc, Palo Alto, Truckee-Donner*
Colorado	Glenwood Springs; Arvada, *Aurora, Colorado Springs, Denver, Fort Collins, Golden, Longmont, Thornton*	
Connecticut	*Hartford, New Haven*	
Delaware		
Florida	Cocoa Beach, Daytona, Jacksonville, Ft. Lauderdale, Gulf Breeze, Manalapan, Monticello, North Miami Beach, Panama City, St. Cloud; *Adventura, Boynton Beach, Deltona, Dunedin, Homestead, Key West, Lakeland, Mariana Key, Miami Beach, Miami Dade County, Palm Bay, Tallahassee, Tampa, St. Petersburg, Winter Park, Winter Springs*	Quincy, New Smyrna Beach; *Jacksonville*
Georgia	Adel, Athens, Quitman; *Alpharetta, Atlanta, Houston County, Rome*	Dalton, North Oaks, South Dungapps; *Sylvester*
Hawaii	*Maui, Waikiki*	
Idaho		

State	Communities/Regions With Wireless Systems	Communities/Regions With Fiber Optic Systems
Illinois	DuPage; *Aurora, Bradley, Chicago, Cook County, Elgin, Rockford, Springfield*	*Naperville, Peru, Princeton, Rochelle, Rock Falls, Rockford*
Indiana	Marion, Linton, Scottsburg; *Beach Grove, Indianapolis, South Bend*	Crawfordsville
Iowa	Cedar Rapids, Des Moines, Marshalltown; *Dubuque*	Cedar Falls; *Opportunity Iowa-state network*
Kansas	Lenexa, Western Kansas; *Johnson City, Kansas City, Leawood, Wichita*	North Kansas City
Kentucky	Lexington, Owensboro; *Louisville*	Murray
Louisiana	Baton Rouge, New Orleans,[†] Vivian, Washington; *Gramblin, Ruston*	*Lafayette*
Maine	Farmington, Franklin County	
Maryland	Allegany County, Ocean City; *Annapolis, Baltimore, Cabin John, Montgomery County, Rockville*	
Massachusetts	Cape Cod, Malden, Nantucket; *Boston, Brockton, Brookline, Cambridge, Newton, Worcester*	*Concord, Taunton*
Michigan	Ferrysburg, Gladstone, Grand Haven, Marquette, Muskegon, Spring Lake; *Ann Arbor, Lansing, Grand Rapids, Oakland County, Ottowa County, St. Ignace, Wasthenaw County*	Cobblestone-Holland
Minnesota	Chaska, Buffalo, Moorhead, Grand Marais; *Minneapolis, St. Louis-Park, St. Paul*	Windom; *FiberFirst*
Mississippi	*Biloxi*	
Missouri	Lewis and Clark County, Nevada, Springfield; *St. Louis*	*North Kansas City*
Montana		
Nebraska	Lincoln	
Nevada	Boulder City; *Las Vegas*	
New Hampshire	Portsmouth	*Hanover*
New Jersey	Hoboken; *Trenton*	
New Mexico	Las Lunas, Rio Rancho; *Albuquerque, Mariposa*	

(continues)

Table 8.1 Status of Municipal Broadband Access in the United States (continued)

State	Communities/Regions With Wireless Systems	Communities/Regions With Fiber Optic Systems
New York	Croton-on-Hudson, Fire Island, Glen Cove, Jamestown; *Buffalo, Glen Falls, New York City, Rochester, Suffolk County*	Ontario County; *Monroe County*
North Carolina	Asheville, Duck, Winston-Salem; *Charlotte, Clayton, Laurinberg*	
North Dakota		
Ohio	Akron, Cuyahoga Falls, Dayton, Dublin; *Cleveland, Dublin, Marietta, Shaker Heights, Van Wert*	*Dover, Butler County, Dublin*
Oklahoma	Oklahoma City; *Norman, Tulsa*	Sallishaw
Oregon	Ashland, Hermiston, Lebanon, Medford, Sandy, Umatilla County; *Corvallis, Easton Oregon Wireless Consortium, Portland*	Douglas County, Independence; The Dalles, Minet-two city project (Monmouth and Independence)
Pennsylvania	Dublin, York County; *Bethlehem, Kutztown, Lansdale, Philadelphia, Pittsburgh, Wilkes-Barre*	Kutztown
Rhode Island	*Providence, statewide service*	
South Carolina	Charleston; *Camden, Columbia*	
South Dakota		
Tennessee	Franklin, Jackson	Bristol, Jackson, Morristown
Texas	Addison, Austin Energy, Corpus Christi, Garland, Granbury, Linden, Southlake; *Abilene, Arlington, Burleson, Dallas, Farmers Branch, Georgetown, Grand Prairie, Houston, Mesquite, San Antonio, Southlake, Tyler*	
Utah	Lindon, Midvale, Murray, Oren, West Valley City; *Utopia: Brigham City, Cedar City, Cedar Hills, Centerville, Layton, Payson, Perry, Riverton, Trememonton*	Provo; *Utopia: Brigham City, Cedar City, Cedar Hills, Centerville, Layton, Payson, Perry, Riverton, Trememonton*
Vermont	Island Pong, Montpelier, Westmore; *Brandon, Brattlesboro, Burlington, Calais, East Montpelier, Grand Isle/ South Hero, Greensboro, Lowell, Marshfield-Plainfield, Ripton, Stamford, West Windsor, Worcester-Middlesex*	Burlington

State	Communities/Regions With Wireless Systems	Communities/Regions With Fiber Optic Systems
Washington	Benton County, Boardman, Feder Way, Kennewick, Pasco, Renton, Spokane, Stevenson, Umatilla and Morrow Counties; *Bellevue, Kirkland*	Chelan County, Clallam County, Douglas County, Grant County, Kitsap County, Marson County, Pend Oreille County; *Seattle*
West Virginia		
Wisconsin	Sun Prairie, Waupaca, Jackson, Shawano; *Madison, Milwaukee, Racine County, Waukesha*	Berseth-Baldwin, Prairie View-Baldwin, Reedsburg
Wyoming	*Cheyenne*	*Rock Springs-Green River*
Virginia	Alexandria, Arlington, Culpepper, Dickenson County; *Fairfax, Manassas*	Bristol; *Danville Lenowisco development district*

Data from http://www.news.com/Municipal-broadband-and-wireless-projects-map/2009–1034_3–5690287.html

*Alaska is in a somewhat unusual situation, in that much of its area is not covered by utility poles. Ultimately, Internet access must come to such regions via satellite, although access to homes and businesses can be through telephone lines or fiber optic cables.

†New Orleans's municipal wireless network was installed during the rebuilding of the city after hurricane Katrina.

Although broadband access may be provided by a municipality, it does not mean the service is free. Rather, the municipalities generally group providing broadband access with other basic services, such as water and electricity. This marks a significant change in the way in which people view Internet access: It has gone from being a luxury to something that everyone should be able to obtain at a reasonable price. In addition, municipalities see Internet access as a way to stimulate an area's economy because it attracts businesses.

There are two ways in which municipalities administer Internet access. In the first situation, the municipality operates as an Internet service provider, providing service directly to businesses and individual consumers. This works well where there is no competition from private broadband service providers. In the second case, the municipality builds the network but then leases the facilities to a private company that administers the service. This second model seems to work best in areas where there is existing competition for broadband customers.

POLITICS AND TECHNOLOGY

Discussing politics with others can be a chancy business. With technology to bring us "up close and personal" to political candidates and leaders, we tend to have firmer opinions than any other time in history. Technology has brought us a great deal of political information, and people continue to argue as to whether this is a good thing.

Television and Politics

Television significantly changed the face of politics. With the advent of news broadcasts and debates, voters were able to see and hear candidates. The first use of television in a political campaign was the "I Like Ike" caricature-filled cartoon that ran as part of Dwight Eisenhower's 1952 Presidential bid.[12]

That first campaign commercial heralded major changes in the way in which the U.S. population viewed candidates and the criteria used to choose a candidate. Most importantly, appearance matters. There has been some speculation, for example, that had there been television in the mid-1800s, Abe Lincoln would never have been elected. He was a tall, gangly man who certainly didn't look like a movie star. Most voters, however, didn't see him; they read his speeches in newspapers, sometimes weeks after they were given. At best, they saw drawings of him that were often caricatures. In an age when campaign travel was by horse, carriage, or train, physical appearance wasn't a significant factor in the choice of an elected official.

The importance of television became very clear after the first televised debate between the candidates for U.S. president in 1960. John F. Kennedy, looking relaxed and handsome, debated Richard M. Nixon, who was tense, poorly dressed, and not as physically attractive. Kennedy stood with his hands crossed in a gentle, open relaxed manner. Nixon's hands, however, looked tense. In addition, Nixon's clothes were ill-fitting, whereas Kennedy wore a well-tailored suit. All in all, Kennedy presented a more competent image that people believed they could trust. It may well be that his performance in the debate cost Nixon the 1960 election.

For approximately 25 years after that first debate, television and politics were a rather sedate marriage. There were debates every presidential election and reporting of political events on network news services. It wasn't exciting, and many people didn't follow the newscasts. However, by 1990 American politicians had discovered that appearing on entertainment television could get them a lot of votes. One of the most famous examples is Bill Clinton's 1992 appearance on the "Arsenio Hall Show." Wearing dark sunglasses and carrying his saxophone, his appearance was a surprise to the audience. The man who would become the United State's 42nd president played "Heartbreak Hotel." Polls indicated that he was losing the election before the television appearance. Afterwards, he rose in the polls and was ultimately victorious in the election. Did the saxophone playing win him the election? Probably not by itself, but there is strong evidence that the television appearance outside of typical political reporting helped to humanize him to the population and provided significant help to his election bid.

Predicting Election Results

Once the results of U.S. national elections were broadcast on live television, predicting the outcome of an election became a favorite pastime of the major television networks. A consortium of networks—the Voter News Service, consisting of ABC, CBS, CNN, Fox News, NBC, and the Associated Press—uses a combi-

---

[12]To see the ad, go to http://www.youtube.com/watch?v=rh6aIkvgyVk

nation of exit polls and computer models to make the predictions, which are then relayed to the viewing public by on-air newscasters.

The multiple time zones spanned by the United States have always made the predictions problematic. Before the 2004 presidential election, networks made the predictions as early as possible, as soon as they had enough exit poll data to use in the computer models. This meant that the winners in eastern states were declared before the polls had closed in the west. If the predictions indicated the winning candidate from just the Eastern and Central time zones—in other words, one candidate had enough votes to win without the results from any other states—voters in the Mountain and Pacific time zones lost their incentive to vote. If the election results were already decided, why bother? Their votes didn't matter anyway.

The situation came to a head during the 2000 presidential election.[13] The Voter News Service predictions came so early for some states that they were broadcast before the polls closed within the state. This had the potential to affect voter turnout—and thus affect the results of the election—in several ways. If supporters of a candidate saw their candidate losing, they could either get discouraged and not vote (making the prediction a fait accompli) or they could be energized to get as many people to the polls as possible in an attempt to turn the election to their candidate. Alternatively, if their candidate was winning, voters could become complacent and not bother to vote at all. In that case, if the opposition turnout was strong, the candidate predicted to win ultimately could lose.

The point of the predictions was to provide interesting information for those following the election on television—not to influence the results of the election. In 2000, Voter News Service kept changing its prediction as to whether George Bush or Al Gore had won the race for president. The problem was centered in Florida, where the predictions changed from one candidate to another several times during the evening hours.

The varying predictions were the result of several technology failures. First, many of the exit poll-takers were unable to phone in their results because Voter News Service's new voice recognition system didn't work properly. Even when workers could finally send in their data, problems with the underlying computer systems prevented the modeling software from working properly. The resulting predictions were often unrelated to the actual vote tallies.

The effects of the inaccurate and badly timed predictions were such that the networks effectively disbanded Voter News Service. During the 2004 presidential election, networks made no predictions until the polls had closed in any given state.

Politics and the Internet

Just about every political candidate—especially those running for state and federal positions—has a Web site. The home page usually includes a flattering picture of the candidate, with links to pages describing endorsements, positions on major issues, campaign videos, podcasts, blogs, and, of course, fund raising.

---

[13]See http://www.baselinemag.com/print_article/0,3668,a=35729,00.asp

How effective are such Web sites? Keep in mind that an Internet user doesn't get to a candidate's Web site by accident. He or she has to make an effort to find the Web site: Either the voter is supporting the candidate already or the voter is interested in finding out more about the candidate. A Web site isn't likely to be visited by someone who is uninterested in the election or not in favor of the candidate.

Nonetheless, fund raising through a Web site is usually very successful, apparently because the Web site attracts those who are supporting the candidate already. In particular, a Web site makes it easier to accept small donations. Registrations at the Web site form a mailing list that a candidate can use to spread his or her message. Assuming that it is possible to assemble a list of marginal voters (those who "might" support a candidate), Web sites can generate maps of the locations of those voters for volunteers to use in planning in-person visits.

The use of the Internet as part of political campaigns has also made it easier for individual citizens to post material either supporting or not supporting a candidate. In some cases, there have been conflicts between the individual and the candidate over who controls such postings. For example, in 2004 Joseph Anthony posted messages supporting the presidential candidacy of Barack Obama on MySpace. When Obama's candidacy became a reality, his aides advised him to take control over what appeared about him on the Internet. Mr. Anthony was not pleased, but ultimately control of the material did revert to the official Obama campaign.[14]

Politicians have also discovered the world of YouTube. In preparation for the 2008 presidential primaries, Republican Party candidates conducted video debates through YouTube videos. First, users submitted their questions by uploading videos. Then, the candidates responded by answering questions in the same manner.[15] There is no valid way to measure how much the debate added to or detracted from a particular candidate's campaign. However, YouTube reached an audience that included people who wouldn't necessarily watch the same debate on television. Allowing users to upload their own questions also made users participants in the debate process, something that isn't possible in a televised debate, where the questions are typically posed by professional newscasters. The increased involvement in the election process meant that some people who might otherwise not have voted went to the polls. Once again, it's difficult if not impossible to determine how many people were affected in this way, but as Americans are fond of saying: "every vote counts."

One question that remains unanswered is whether use of the Internet will increase the participation of young people in the political process. The research we have considered earlier in this book suggests that the largest proportion of Internet users are young; exit polls have indicated an increase in the percentage of young people after the year 2000. There's no way to be sure that the Internet is causing the increase, but logic suggests that it is responsible for at least some additional voters.

---

[14]For details, see http://www.boston.com/news/nation/articles/2007/05/10/internet_and_politics_an_uneasy_fit/
[15]To see for yourself, go to http://www.youtube.com/republicandebate

There is a potential downside to Internet use in politics: Because a user must choose to visit a Web site—and make the effort to do so—the user isn't exposed to as much information from opposing candidates as he or she would be if political information was coming from newspapers, radio, or television. Although hearing or reading about positions with which we don't agree can be uncomfortable or even make us angry, it is important that we do so to put our own positions in perspective.

TECHNOLOGY AND WARFARE

There is a theory that wars spur the development of technology. Certainly, warfare has provided a testing ground for new technologies and, at the very least, sped up the development of specific technologies. In this section we will begin by looking at some historical weapons technologies. Then we will look at the technology that was used and developed for three major armed conflicts, and, as we talk about the war in Iraq, we will consider some current military technology and the problems that have arisen with its use.

Historical Weapons Technologies

Just as humans have a long history of creating devices to help with computation, they also have a history of creating weapons. Technological weapons date back to the 14th century, although the use of chemical weapons dates back to the 5th century, when the Spartans used sulfur fumes against their enemies.

A summary of early weapons technologies can be found in **Table 8.2**. Notice that most weapons focus on uses of gunpowder and better guns. Combatants used poison gas widely in World War I, but the results were so horrifying that the world as a whole decided to ban their use. The ban was so well accepted that gas was not used in combat in World War II.[16]

World War II

World War II was the first war to make extensive use of technology, including aircraft, radar computing machinery, rocketry, and atomic bombs. It many ways it also contributed more than any other war to the advancement of technology. For example, World War II saw the development of duct tape, which soldiers used to hold broken parts together. It entered civilian markets only after the war, when heating and cooling installers used it to hold ducts together.

Aircraft Aircraft were used at the very end of World War I, but they weren't seen as providing a strategic advantage. In fact, some pundits of that era believed that aircraft would put an end to war because they allowed combatants to gain an aerial view of their opponents' positions, totally eliminating the element of surprise in an attack. Instead, by the time World War II began, air power was an accepted means of dropping men and bombs on an enemy.

---

[16]This is not to imply that gas hasn't been used since World War I. The Germans certainly used poison gas in their concentration camps. Iraq used gas on the Kurds in the 1990s.

Table 8.2 A Brief History of Early Weapons Technology

Year(s)	Weapon
5th century	Spartans use sulfur fumes to poison enemies
14th century	Development of gunpowder in China
1324	First reliable reference to a gun
1326	First reliable reference to a cannon
1380s	Guns spread across Europe
Early 15th century	Development of the matchlock gun
1498	Rifling is developed
1509	Development of the wheel lock
1530	The first guns to use rifling
1630	Flintlock guns appear
1726	First description of a pressure-activated land mine
1820s	Percussion-cap guns appear
1830	Back action guns appear
1835	Colt makes its first revolver
1840	Guns with pin-fire cartridges appear
1850	Shotguns appear
1859	Rim-fire cartridges appear
1860	A patent is issued for the repeating rifle, issuing in the era of breech-loading guns
1861	The Gatling gun, the precursor to the machine gun, appears; mines that explode on contact are used by the U.S. Navy
1869	The center-fire cartridge appears
1871	A revolver that uses cartridges appears
1873	The Winchester rifle appears
1877	The double-action revolver appears
1879	The first machine gun appears. It fires 10,000 rounds in 27 minutes.
1892	The first automatic weapons appear
1909	Germany develops the first antiaircraft guns. However, because the use of aircraft wasn't considered a viable war tactic, the guns weren't produced for World War I.
1915–1916	World War I is fought, including the first widespread use of poison gas, which kills 100,000 and injures another 900,000.*
1915	The first tank running on caterpillar treads appears (**Figure 8.3**); Britain uses an antiaircraft gun to bring down an airplane for the first time.

*The use of chemical weapons was banned in 1928 by the Geneva Protocol.

Figure 8.3
The first tank
("Little Willie").

The propeller-driven fighters of World War II have become the stuff of legend. Planes such as the British single-man Spitfire (**Figure 8.4**) and the U.S. carrier-based Corsair were used in the plane-to-plane battles known as dogfights against the Japanese Aichi D3A that have been romanticized in film and television since the middle of the war.

World War II combatants also used bomber aircraft such as the B-17 Flying Fortress (**Figure 8.5**). Although we might consider such a propeller-driven plane primitive by today's standards, it was able to sustain a great deal of damage—such as losing all its landing gear—but still land safely or float for a time after ditching in the water. It had a long range and could fly higher than many fighter aircraft; it also carried several guns that could be used to defend the plane while in flight.

Figure 8.4
British World War II
fighter plane, the Spitfire.

Figure 8.5
The U.S. B-17 bomber.

World War II spurred the development of jet aircraft, although many were experimental and never saw combat. The first jet used by the U.S. Army Air Force was the P-80 Shooting Star. First flown in 1943, it saw only limited action at the end of World War II but was used extensively during the Korean War. Many people would argue that the civilian adoption of jet aircraft was accelerated by the work done to develop jets for use in warfare.

Computing   A great deal of early work on the development of computing was spurred by the needs of World War II, in particular for computing the angle at which a large gun should be aimed to reach a particular target. In parallel, the code-breaking research at Bletchley Park in England lead to the development of the Enigma Machine and to the early computers named Colossus.[17] Colossus, which remained a British national secret until 2000, was not only able to break codes that the mechanical machines could not but also to predict some of Hitler's actions. The British credit Colossus for shortening the war by two years.

Rocketry   World War II also saw the development of rocketry into a usable form. Development of a military rocket began in 1936. The development effort was lead by Wernher von Braun (**Figure 8.6**), who immigrated to the United States after the war and became instrumental in the U.S. space program.[18] By 1941, the V-2 rocket was capable of delivering its payload, a bomb that could not be defeated by electronic defenses, antiaircraft guns, or fighter aircraft. To defend against the rockets, the Allies needed to destroy the launching pads or trick the Germans into aiming them somewhere that would do little damage.

The V-2 rocket was a superior weapons technology to that of the Allies, but it did not turn the war in Germany's favor, for several reasons:

- The Allied advance pushed the launchers back so far that the range of the rockets was not sufficient to reach targets in Allied territory.
- The V-2 could not explode in the air. Instead, it had to burrow in the ground before it detonated, which decreased its effectiveness.
- The V-2 had only a primitive guidance system, and targeting it to specific destinations was impossible.
- The V-2 was very expensive. Each one cost almost as much as a four-engine bomber. The bombers could fly farther, faster, carry multiple bombs, and were somewhat more accurate in hitting their targets.

The V-2 did have a major psychological effect, primarily because it was silent. Nonetheless, it was not enough to tip the war in Germany's favor.

---

[17]For details, see http://news.independent.co.uk/uk/this_britain/article259133.ece
[18]There was a saying floating around the U.S. space program for a time that explained why the U.S. space program ultimately pulled ahead of the Russians: "Our Germans are better than your Germans." Although there is little documented proof, it appears that most German rocket scientists went to the United States or Russia after the war to continue their work. The quality of the scientists each government was able to employ made a significant difference in the development of space technology.

Figure 8.6
Wernher von Braun.

Reporting the War Before World War II, technology had a very minor role in bringing news of a war to people back home. Most of those not directly involved in the war received their news from newspapers that got their stories from correspondents in the battle zones. News was often weeks out of date. Reporters during World War II had better tools at their disposal, however.

World War II reporters were able to use transatlantic cables to telegraph information home; the same cables could be used to telephone news across the Atlantic. Correspondents could use still photography, such as the scene in **Figure 8.7,** and video captured on film to record images. News was broadcast over the radio as well as printed in newspapers. Film images were assembled into newsreels that were typically shown in movie theaters before the feature. Although they appeared weeks after the events depicted, the newsreels represented the first time those at home were able to see video of the war in action.

Atomic Energy World War II also saw the development of the world's most devastating weapons technology: the atomic bomb. As early as 1939, Albert Einstein was aware of German efforts to enrich uranium, a resource that was essential for an atomic bomb. He communicated his concerns to President Franklin Roosevelt. Shortly thereafter, the United States established The Manhattan Project with the express goal of developing an atomic weapon before Germany could do so.

From 1939 to 1945, scientists spent more than $2 billion to create the first working atomic bomb. It was what was known as a "gravity bomb," which had to be dropped from a bomber. The first test, in the open air on July 16, 1945 in New Mexico, produced the mushroom cloud that was to become a symbol of the horror of atomic weapons.

Figure 8.7
A World War II battlefield photograph: the first flag raised on Iwo Jima.

Atomic bombs have been dropped twice in warfare, both times by the United States: on August 6, 1945 on Hiroshima, Japan and on August 11, 1945 on Nagasaki, Japan. The results were devastating. More than 120,000 people were killed by the initial explosions, and many thousands more died from radiation sickness over the ensuing decades. The technology of the atomic bomb and its potential for destruction laid the foundation for international relations in the post war era. The Cold War remained "cold" because of the fear of nuclear weapons; many of the tensions between nations in the first part of the 21st century stem from the proliferation of nuclear weapons and the fears that the smaller countries that have them may one day use them.

The Vietnam War

By the time the United States became involved in the Vietnam conflict, jets were a staple of warfare, used primarily to drop bombs. Perhaps the most well known, the McDonnell Douglas F-4 Phantom (**Figure 8.8**), was also used as a fighter aircraft. Older types of aircraft, such as the Lockheed C-130 Hercules turbo prop (**Figure 8.9**), were used as cargo carriers. Although helicopters were used during the Korean War, they played a more important role in Vietnam. The CH-47 Chinook (**Figure 8.10**), for example, was used to carry men and materiel throughout combat areas.

The Vietnam War also saw the development and use of laser-guided bombs (what we now call "smart bombs"). Previously, "gravity bombs" had been simply dropped from aircraft and let fall where they would; needless to say, they couldn't be targeted very precisely and as often as not landed in unintended locations, missing the target the military planned to hit. Laser-guided weaponry requires that the target be marked by a targeting indicator; then the laser can use the mark to home the bomb on the target. Because the laser has to "see" the targeting mark, laser-guided bombs can't be used in bad weather. However, they are far more precise than gravity bombs.

Figure 8.8
The F-4 Phantom fighter/bomber, used in the Vietnam War.

Figure 8.9
The Lockheed C-130 Hercules turbo prop, used as a cargo carrier during the Vietnam War.

Figure 8.10
The CH-47 Chinook helicopter, which was used extensively in the Vietnam War.

The Vietnam War also saw the development of some technologies that we take for granted today:

- Night-vision goggles
- Global positioning satellites, or GPS
- Seismic sensors that track the movements of people and animals[19]

---

[19]For details on how seismic sensors are used to help elephant populations, see http://www.bio-medicine.org/biology-news/Vietnam-war-technology-could-aid-elephant-conservation-1080-1/

Although beneficial technology did come out of the Vietnam War, there were also some technologies that were horrifying. For example, U.S. troops used the herbicide Agent Orange to destroy foliage that was hiding North Vietnamese troops. Not only did the herbicide destroy the countryside, but the chemicals in Agent Orange and other similar compounds break down into dioxins, which are known to increase the risks for cancer and genetic defects. In addition, the effects of the deforestation remain today, including the losses of topsoil that occurred when rains washed down deforested hillsides.

U.S. forces also used a compound called Napalm. Originally developed in World War II, Napalm was an incendiary gel that was used to increase the effectiveness of flame-throwers. It also could be used as a bomb. When dropped, a Napalm bomb burned so quickly and so hot that it removed oxygen from the surrounding air, leaving carbon dioxide behind. Anyone in the path of such a device could be burned severely if the burning gel adhered to skin or clothing; a person could also be suffocated by the carbon dioxide.

Reporting the War Technology of a different sort played a major role in public perception—and perhaps the outcome—of the Vietnam War. Major news agencies equipped reporters with the newly developed portable videocassette recorders and sent them to the front lines with the troops.[20] For the first time, audiences at home were able to see the action of a war—both the good and the bad—in close to real time. The action in previous wars was far away from those at home and was often "sanitized" by the military and news services to keep the horror of the war from ordinary citizens. Seeing what actually went on during an armed conflict was at least a part of what turned public opinion in the United States against the Vietnam War.

The War in Iraq

Fast forward some 35 years, and the war in Vietnam becomes the war in Iraq. The United States is once more involved in a conflict far from its own borders that has no easily foreseeable end. However, in terms of technology the two wars shared only a little. The military that serve in Iraq are infused with more technology than any other force in history.

Probably the biggest change in U.S. military technology involves communications: Today's army is networked, linked by satellites. Ground vehicles such as Humvees are equipped with satellite transponders that let the location of vehicles in the field be monitored at all times. At headquarters the movements of troops and equipment are plotted by computers on electronic maps. Communications in soldier's helmets identify friendly forces (although they can't pinpoint the location of enemy forces).[21] An increasing use of robotic aircraft and ground vehicles provides information to headquarters and troops in the field through wireless communications.

---

[20]The cameras carried by Vietnam era newsmen used the ¾″ U-Matic tape format (the format of the first videocassettes) and therefore were rather large, bulky, and heavy.

[21]Networking also keeps soldiers in contact with those at home. In previous wars, families relied on letters, the occasional (and often random) appearance in a news video, and rare telephone calls for contact. Today, soldiers in combat zones use e-mail, text chat, and video chat to contact families. Conversations with home are much closer to "real time" than they have ever been.

For the Iraq war the state-of-the-art in infantry equipment is Land Warrior. As you can see in **Figure 8.11**, the outfit consists of a weapon, protective clothing and armor, and technology integrated into the helmet. The weapon—based on the M16 rifle—includes the basic weapon's firing capabilities and additional technologies such as a daylight video sight, thermal sight, and a multifunction laser sight to supply range and direction information. The helmet has an integrated organic light-emitting diode display (an LED that uses organic compounds for pixels) that shows the soldier digital maps and the locations of friendly forces. Theoretically, that information will be relayed to the weapon, letting the soldier see and fire around corners. The Land Warrior also provides GPS capabilities and a radio. All the technology is connected to a small wearable computer powered by a battery.[22]

In addition, spy satellites provide detailed images of enemy troops and weapons emplacements on the ground. Satellite-guided bombs provide even more precise targeting than laser-guided weapons. Several technologies have also been pilot tested in the war zone:

- Remote-controlled drone aircraft
- Remote-controlled ground vehicles[23]
- Nanotechnology as part of communications and computing systems
- Peer-to-peer communications (as opposed to sending all communications through a central server)

However, given the nature of the conflict in Iraq, superior technology hasn't been enough. Much of what occurrs in Iraq depends on the personal interactions between ground forces and the citizens of the country. Technology can't build the trust needed to get local leaders to cooperate with troops. Doing so requires old-fashioned face-to-face contact.[24]

Figure 8.11
A soldier wearing the Land Warrior equipment used in Iraq.

[22]For an analysis of the Land Warrior system, see http://blog.wired.com/defense/2007/09/when-the-soldie.html
[23]For details, see http://www.news.com/8301-10784_3-9827572-7.html?part=rss&subj=news&tag=2547-1_3-0-5
[24]For more information, see http://www.wired.com/politics/security/magazine/15-12/ff_futurewar

Advances in technology during the Iraq War, especially improvements in personal body armor, have increased the survival rate of injured solders. One effect of this increased survival rate is a major need for prosthetic limbs. Therefore one collateral advancement to come from the Iraq War is a jump forward in the technology of prosthetics. With the development of the C-leg[25] and i-Limb[26] prosthetics, doctors and scientists have come very close to providing a bionic limb replacement.

Reporting the War During the early days of the Iraq War, journalists were "embedded" with the troops. In other words, they were at the front lines with combat units.[27] The technology for instantaneous transmission of information was certainly available: Satellite uplinks could be powered by small solar panels if necessary. However, the embedding gave journalists access to some information that should *not* be made public (or that the military did not want made public), such as details of troop movements or classified weapons. Embedded journalists therefore had to agree to the following restrictions:

- Military actions could be described only in general terms.
- No information about future missions could be divulged.
- If the military unit commander declared a "blackout," no satellite transmissions were allowed from the field. This was to be used in cases where the satellite uplink potentially could reveal the location of the military unit.

The result was that the coverage of the war in Iraq has been less detailed than that of the war in Vietnam, where there were few restrictions on what newspeople could report, despite the advances to communications technology in the 35 intervening years.

---

[25] For details, see http://www.engadget.com/2005/06/22/let-your-c-legs-do-the-walking/
[26] See http://www.usatoday.com/tech/news/techinnovations/2007-07-19-bionic-hand-amputee_N.htm
[27] Yes, embedding journalists with front-line troops meant that more journalists were killed than in previous wars. However, this had little to do with technology but rather with where the journalists were.

WHERE WE'VE BEEN

Governments run on data. Before the introduction of computer systems, data were kept in paper files. The difficulty of maintaining and searching such files has led governments to be early adopters of technology. Some governments have embraced electronic databases and the Internet as part of a national policy; others have added electronics in a more piecemeal fashion.

Technology has made it easier for governments to store information about their citizens. It has also made it easier for governments to gather information using devices that intercept electronic communications. Major issues remain unsettled concerning access to government files and ensuring the privacy of the data gathered.

Throughout history, warfare has used increasingly sophisticated technology. "Modern" forces strive to have the best weapons and the best armor (for both men and vehicles). However, the best technology may not be enough, because many times interactions with people are as important as firepower.

Many technologies have either originated in response to military needs or received increased development support during wartime. Computing, in particular, benefited from the needs of World War II: Early computers were used to aim artillery and to crack enemy codes.

THINGS TO THINK ABOUT

1. Have you used the Internet to find political information? If so, what type of information did you look for and why did you seek it? Did what you found satisfy you? Why or why not? If you have not used the Internet to find political information, what would entice you to do so? What information would you like to see available?

2. The development of Colossus and other computing machines during World War II supports the theory that wars spur on the advancement of technology. Research the technology of the Korean War (the "peace action," as it was called officially). What technologies, if any, were first used in combat in this war? Does this support or contradict the theory?

3. Journalists reporting the Iraq War are a given more access to the fighting and military units than in any other war. However, as you read in this chapter, there are some restrictions placed on what they could report. Should the journalists agree to these restrictions? Why or why not? How do we reconcile the conflict between freedom of the press and security for military personnel in the field?

4. As you read in this chapter, some local governments are providing free or low-cost broadband Internet access to their citizens. Should governments be in the business of providing Internet access? Why or why not?

5. The use of television in politics has made it increasingly important that candidates for office look good, move well, and dress well. Television has also made it easier for more people to see and hear candidates. In your opinion, has the use of television in the political process had a positive or negative effect? Why?

WHERE TO GO FROM HERE

"Agent Orange: Information for Veterans, Their Families and Others About VA Health Care Programs Related to Agent Orange." U.S. Department of Defense. *http://www1.va.gov/agentorange/*

Hastings, David. "Politics, Elections and the Media." *http://www.transdiffusion.org/emc/insidetv/*

Kennedy, Gregory P. Vengeance Weapon 2: The V-2 Guided Missile. Washington, D.C.: Smithsonian Institution Press, 1983.

"Municipal Broadband Networks." *Light Reading*, October 04, 2005. *http://www.lightreading.com/document.asp?doc_id=80313*

Museum of Broadcast Communication. "Political Processes and Television." *http://www.museum.tv/archives/etv/P/htmlP/politicalpro/politicalpro.htm*

Nagourney, Adam. "Politics Faces Sweeping Change via the Web." The New York Times, April 2, 2006. *http://www.nytimes.com/2006/04/02/washington/02campaign.html?ex=1301630400&en=d566826d88d5f5cf&ei=5088*

"Napalm." GlobalSecurity.org. *http://www.globalsecurity.org/military/systems/munitions/napalm.htm*

Oates, Sarah, Diana Owen, and Rachel Gibson. The Internet and Politics: Citizens Voters and Activists. New York: Routledge, 2006.

Pae, Peter. "The Autonomous Warbird." Los Angeles Times, November 27, 2007. *http://www.latimes.com/entertainment/news/business/newsletter/la-fi-robotplane27nov27,1,2515936.story?ctrack=1&cset=true*

"Pros and Cons of Embedded Journalism." *http://www.wired.com/politics/security/magazine/15-12/ff_futurewar*

Talbot, David. "How Technology Failed in Iraq." Technology Review, MIT. *http://www.technologyreview.com/Infotech/13893/?a=f*

von Braun, Wernher, Frederick I. Ordway III, and Fred Durant. Space Travel: A History: An Update of History of Rocketry and Space Travel (4th ed.). New York: HarperCollins, 1985.

Ward, Mark. "Colossus Cracks Codes Once More." BBC News. *http://news.bbc.co.uk/1/hi/technology/7094881.stm*

Weapon: A Visual History of Arms and Armor. New York: DK Publishing, 2006.

Children, Education, and Libraries 9

WHERE WE'RE GOING

In this chapter we will

- Learn when children are capable of using computers.
- Discuss whether young children should use technology and, if so, at what age.
- Discover the types of software available for children of varying ages.
- Learn about the use of technology in the education of children.
- Examine how libraries have embraced technology.
- Show the availability of library materials in electronic form.

INTRODUCTION

One of the most controversial aspects of modern technology is the use of electronics by children. The issues range far beyond the dangers to children from sexual predators on the Internet, including when it is best to introduce children to information technology, how much "screen time" children should have, and whether information technology is appropriate for use in the classroom. Although educators have been researching the impact of technology on children, libraries, long the repositories of educational materials of all types, have been changing to embrace the new technology, In this chapter we will discuss how children and technology fit together and then look at what has happened to our libraries.

TECHNOLOGY AND CHILDREN

In an article describing the behavior of kindergartners when using technology, Hyun and Davis make what is really a profound statement about technology and children: "Over the past two decades, researchers investigating young children and computer technology moved from questioning whether computers can help young children learn to suggesting *how* educators and parents can best use computers to maximize learning."[1] Computers are now an accepted part of the home and learning environments for children, although there are still those who are opposed to the use of technology by the very youngest. In this section we examine some of the issues surrounding the use of technology—particularly computer technology—by children. Even with all the research that has gone into the subject, there are still no clear answers.

Technology and Child Development

When are kids capable of using technology? Even if they are mentally and physically capable, should they use technology at a given age? Some of the answers can be found in research that identifies what children actually do.

A survey conducted by Princeton Survey Research for the Kaiser Family Foundation found that children were using technology at a very young age. **Table 9.1** shows that most children have used a variety of electronic technologies by the time they are 3 years old. A child is physically able to use a mouse by age two, but many children are not cognitively ready to make the connection between moving the mouse and on-screen cursor movements at that young age.

Many educators, psychologists, and parents are concerned more about the amount of time children spend using technology than early use of technology itself. The Kaiser Family Foundation survey found, for example, that children younger than the age of six watch an average of 90 minutes of television a day. However, about a third of the children were growing up in households where the television was on "always" or "most of the time." In addition, just over a third of the children had televisions in their bedrooms. Television also had a negative relationship with time spent reading.

The same survey indicated that in a typical day, children spent an average of about an hour using the computer. Interestingly, children who were heavy computer users (more than an hour a day) also played outside longer (2½ hours a day vs. light computer users at two hours a day).[2] In contrast to television, however, less than 10% of the children had computers in their bedrooms and a mere 3% had Internet access in their bedrooms.

---

[1] Hyun, Eunsook, and Genevieve Davis. "Kindergartners' Conversations in a Computer-Based Technology Classroom." *Communication Education* 54 (April 2005): 118–135.
[2] There is a saying in statistics that "correlation does not imply causation." In other words, just because two things are closely related does not mean that one causes the other; often a third factor causes both effects. In the case of children who are heavy technology users being more physically active than those who use technology less, it is likely that one doesn't cause the other but that some third factor is at work. It may be, for example, that children who are heavy technology users are more engaged in all types of activities than those who don't use technology as much.

Table 9.1 Survey Data Showing the use of Technology by Children
Ages 1 to 5 Years

Type of Technology	Average Age
Watch television	1 y, 2 mo
Turn on television without help	2 y, 1 mo
Use a remote to change television channels	2 y, 5 mo
Use a computer sitting on an adult's lap	2 y, 8 mo
Use a computer without sitting on an adult's lap	3 y, 6 mo
Turn on a computer without help	3 y, 5 mo
Use a mouse to point and click	3 y, 4 mo
Put a CD-ROM into the computer	3 y, 8 mo
Look at a child's Web site	3 y, 4 mo
Ask to go to a specific Web site	3 y, 8 mo
Go to a specific Web site without help*	4 y
Send e-mail with help	3 y, 9 mo
Send e-mail without help	3 y, 2 mo
Play computer games	3 y
Play video games using a game console	3 y, 7 mo
Play video games on a handheld device	3 y, 8 mo

* The survey doesn't indicate whether the child typed a URL or used a bookmark set up by
an adult. From Ridout, Victoria J., Elizabeth A. Vandewater, and Ellen A. Martella. "Zero
to Six: Electronic Media in the Lives of Infants, Toddlers, and Preschoolers." Kaiser Family
Foundation Research Report, 2003.

The National Association for the Education of Young Children (NAEYC)
offers recommendations on the use of technology by young children. For children
ages three through eight years, the NAEYC takes the position that technology can
be an effective learning tool and that it is up to a qualified classroom teacher to
decide what is appropriate.[3] Nonetheless, the NAEYC views technology as only one
method in a portfolio of techniques that can be used when teaching this age group.

In contrast, some in the education field believe that technology has no place
in the lives of young children. For example, Healy[4] writes the following:

> Today, I believe computers should not be introduced until at least the
> second grade for normally developing children. Research says that before
> age 7, children's brains experience rapid growth and change while they
> work toward more abstract ways of processing information.

---

[3] http://www.naeyc.org/about/positions/PSTECH98.asp
[4] Healy, Jane M. "Young Children Don't Need Computers." *Education Digest*, January 2004.

Joining Healy are those who believe that no child under the age of two should use technology (computers, television, and so on) at all. In his article opposing the use of technology with toddlers, Miller[5] states the following:

> Computer play is different from the kind of imaginative, child-initiated play that has long been regarded as the foundation of the early childhood curriculum. Educators report that many children that are adept at playing video games, pushing buttons, and operating a mouse show an alarming lack of imagination.

The American Academy of Pediatricians, while recognizing that electronic media can be an effective tool for educating older children, recommends that children under the age of two should not watch television at all (much less use a computer).[6] The trends in popular media, however, seem to fly in the face of this advice. For example, the Baby Einstein series of DVDs[7] sells extremely well. BabyFirst TV, launched in 2006, is a channel that offers "24/7 DVD-quality programs with a unique parent co-viewing experience...."[8] BabyFirst TV states that its approach to providing television for infants and toddlers is "created by leading experts in child development, education and psychology and provides a safe and enjoyable experience for both parents and babies." The channel does offer printed subtitles for many of their programs, supposedly "making the experience as educational and engaging as reading a book."

So, whom should we believe? Does technology use help or harm young children? Heavy use of technology by children is still such a new phenomenon that we can't be sure. It may take an entire generation of children growing up with technology before we can see the effects of technology on the adults that these children eventually become.

Software for Kids

It wasn't long after computers entered homes that software directed at children began to appear. The younger the child, the more likely the software was "educational." How much software is available for a particular age group has changed over time, depending a great deal on the current thinking of what is appropriate computer use for the age group.

Infant and Toddler Software During the late 1990s and early 2000s there was a great deal of software released for very young children. These titles, designed for infants as young as nine months, required the child to sit on an adult's lap and press the keyboard to instruct the software to change something on the screen. For the most part, this software is designed to teach children that pressing a key affects the computer, and little else. The images tend to include letters, numbers, shapes, and bright colors.

---

[5] Miller, Edward. "Fighting Technology for Toddlers." Education Digest, November 2005.
[6] http://aappolicy.aappublications.org/cgi/content/full/pediatrics;104/2/341
[7] See http://www.babyeinstein.com/
[8] See their Web site at http://www.babyfirsttv.com/

By the latter part of the first decade of this century, the thinking of educators and software developers toward software for the very young had changed. Many of the programs—for example, JumpStart Baby and Sesame Street Baby—were discontinued, although educational software for older children from the same developers remains on the market. At the time this book was written, there was very little up-to-date software for infants on the market.[9]

The thinking behind software for the very young was that it would help teach cause and effect: You hit the keyboard and something happens. Because the effect is consistent—a picture appearing on the screen—the child would learn that one action causes the other. But research into child development indicates that infants aren't able developmentally to make the connection between cause and effect. Therefore the software can't possibly do what it was intended to do.

Preschool Through Second Grade Today's preschool and kindergarten curriculum is far more academic than that of even 20 years ago. Children are expected to leave preschool able to recognize the letters of the alphabet, count to at least 20, and write their first names. Kindergarten begins the formal teaching of reading and writing; a child who does not read by the end of the kindergarten year is at a distinct disadvantage when entering first grade. Therefore much of the current software aimed at this age group teaches basic literacy concepts.

To help engage the children, many programs are based on children's television shows. One series, for example, takes off from a series named "Arthur." (Arthur is an aardvark with a short nose and all his friends are animals of various species, all of whom behave pretty much like human children.) The second-grade program teaches addition and subtraction of double- and triple-digit numbers, map reading, and ancient history (Africa, Egypt, Native Americans, Ancient Greece, Ancient Rome). Kids can learn phonics and hone their reading skills with characters from the show "Clifford the Big Red Dog."

Computer use is virtually a given with children in the primary grades. School districts have pursued the goal of placing computers in classrooms and establishing computer labs since the 1980s. Even if a child doesn't have a computer at home, it is highly likely that he or she will be exposed to one at school and may even participate in "technology education" units. A parent who was opposed to his or her child's use of technology in school would need to make a special request to the school to have that child exempted from technology requirements.

Because computer use by primary-grade children is almost certain, the question remains about the best role of children's software in their educations. From kindergarten forward, educators seem to believe that computer use has a place in the classroom as an enrichment tool, as long as it can be integrated into the curriculum.[10] One of the roadblocks to true curriculum integration, however, is that traditionally teachers have not been trained to use technology. Some find it puzzling or frightening; in some cases, the students may know more about technology than the teachers.

---

[9]You can certainly purchase older titles from a number of online sources. However, the software hasn't been updated in some time. For example, most developers of early childhood software for the Macintosh didn't bother to port their products from OS 9 to OS X.

[10]See "Integrating Technology in Classrooms: We Have Met the Enemy and He is Us." *Association for Educational Communications and Technology*, 27th. Chicago, IL, October 19–23, 2004.

Elementary School Beyond Second Grade In first and second grade the curriculum focuses on language arts skills, just as it did in kindergarten. Beyond that point, however, emphasis switches to using language arts skills to learn material in other subjects. Much of the software aimed at the 8- to 12-year-old age group therefore focuses on topics other than language arts.

Some of the software packages are considered to be very effective teaching tools. For example, the series featuring the fictional spy Carmen Sandiego (Where in the World is Carmen Sandiego?, Where in the U.S.A. is Carmen Sandiego?, and so on) teaches geography in the guise of a game that tracks the title character.[11] The Math Blaster series (for example, Math Blaster 1: In Search of Spot) provides math drills and practice in a game format.[12]

Beyond Elementary School As children grow beyond elementary school, use of educational technology begins to change. There is a significant drop-off in the amount of "educational software" available. Instead, students tend to use productivity software such as word processors, presentation packages, and e-mail.

The biggest change for students is use of the Internet for research. The vast collection of information available through the Web means that it is possible to do a great deal of research for school papers without ever visiting a library. Although about 20% of K–12 students do use a library for school-related research, many now turn to the Internet first and often express frustration when material they need is not available online.[13]

Kids on the Web

Children use the Internet, either for pleasure or research, at home or at school. It's something that educators and parents have to deal with as a reality. The other reality is that the Internet can be a dangerous place for children. There are two strategies for protecting children on the Internet: restrict them to Web sites designed for children and filter the URLs that a child can access.

Game Sites for Kids A number of Web sites have been developed specifically for children by television networks and toy manufacturers. For the very young, there are sites such as pbskids.org and nickjr.com. Elementary and middle school children are drawn to sites with more of an edge, like that run by Cartoon Network. These sites are free but do advertise in some way. The Lego Web site, for example, provides games and cartoons but also advertises Lego products. Blatant advertising also appears at postopia.com, where the characters in the games are pieces of Post cereal. As with broadcast television, advertising is the price users pay for free programming. However, whether children should be exposed to such commercialization is still an unanswered question.

---

[11] See http://www.gamefaqs.com/computer/apple2/review/R12493.html for a review.

[12] For a review see http://www.superkids.com/aweb/pages/reviews/math1/blaster1/merge.shtml

[13] For many years I did online tutoring for AOL in their "Homework Help" area. (The live help was discontinued in the mid-2000s.) Often, it was difficult to get a younger student to understand, for example, that a newspaper article from 1940 couldn't be found online and that a trip to a library was necessary.

To avoid advertising, parents can purchase memberships in game sites. For example, Disney's ToonTown appears to be aimed at third grade and above (based on the reading level required). It offers some free play but restricts full access to those who subscribe. At $9.95 a month or $79.95 a year, many parents may believe that the fee is excessive, however.[14]

Homework Sites Not all children's Web sites are for playing games. There are a number of "homework help" sites that provide lessons in subjects taught in grades 1 through 12. Many sites charge for access to lessons and tutorials although factmonster.com is free. Homeworkhelp.com charges up to $175 a year ($30 for a single month) to use any of 3,000 self-paced learning modules for grades 4 through 12. Tutor.com provides "live" help by letting students interact with a tutor in real time using both chat and an online interactive whiteboard, although at $29 an hour most parents would consider it rather expensive.

How ethical is the use of homework help sites? Certainly it depends on exactly what type of help is being provided. Most free sites provide prewritten lessons and reference materials. The chance that a student will plagiarize from such a site is very real but not high given that the lessons teach concepts rather than solve specific problems. However, live tutoring sites, just like with homework help given by a parent, run the risk of doing too much of a student's work.

Of considerably more dubious ethics are the sites that provide prewritten term papers. Some are free. Students can upload papers they want to share and download papers they need. Because the papers are written by students, they are less likely to make a K–12 teacher suspicious than papers written by experts. College students, however, can request custom-written research papers for a fee.

The problem of the availability of prewritten term papers is only a partial source of plagiarism issues that have arisen with the use of Web sites for homework. Rather than submit someone else's paper as their own, some students copy paragraphs from Web sites to paste into their work without citing the source or indicating that someone else wrote the material. In some cases it is very clear that a student has copied from the Web, especially when the copied material has a writing style different from that of the student. However, identifying downloaded papers can be very difficult, especially if the entire paper was written by another student.

To help with the problem, Web sites such as Turnitin[15] have arisen. Schools contract with Turnitin to produce "Originality" reports for papers submitted by students. Turnitin compares a submitted paper to its own database of submitted papers, current and previous Web sites, and journal/magazine articles. The report indicates how similar the submitted paper is to existing materials. Despite the range of sources used to make comparisons, Turnitin doesn't claim to be able to catch all plagiarism or that there will be no false positives. However, the site provides a significant help for teachers who are attempting to ensure that students do their own work.

---

[14]The $9.95 a month adds up to almost $120 a year, when paid monthly. Contrast that with the $40 a year adults pay to use pogo.com, a game Web site designed for those over age 13.
[15]www.turnitin.com

Content Filtering The Web is largely an unregulated playground. With the exception of child pornography, there is little that can't be posted to a Web site. Most content-filtering software attempts to prevent children from accessing inappropriate Web sites. Some Web site administrators make that difficult, however. Consider the "whitehouse" domain name. www.whitehouse.gov is the gateway to the U.S. government's official Web site. You might want to send a student there to see who his or her state representative is or to look up pending legislation. At one time, however, the Web site www.whitehouse.com was a brothel. The .com suffix is so ingrained in most Web users that people type it almost unconsciously. It was very easy for a child to accidentally end up at the wrong site by accident, which was exactly the intent of the whitehouse.com site administrators.[16]

Many parents have installed Web-filtering software in an attempt to block inappropriate Internet activities. An Internet filter features the ability to block and/or monitor chat, newsgroups, instant messaging, peer-to-peer networking, FTP, and e-mail. Users can also establish lists of blocked Web sites and create sets of keywords that can be used to flag undesirable sites. Most filtering packages also prepare a history of where a user of the filtered computer has been on the Web and will send a report to another machine. Some also enforce time limits on the use of an Internet browser. Although software filters can't block every Web site a parent might want to prevent a child from accessing—the Web is simply too dynamic for that—such software can go a long way to keeping children from inappropriate Internet access.

TECHNOLOGY AND LIBRARIES

Somehow libraries have earned a reputation for being cavernous halls of books run by middle-aged spinsters with buns and sensible shoes whose only function was to ensure quiet. If that image were ever true, it certainly isn't today. The use of technology in libraries is as significant as in any industry you can name. In this section we will look at how things were done in libraries before the introduction of technology and at how technology has changed the way libraries achieve their goals of organizing and providing information.

How Libraries Have Adopted Information Technology

Libraries were early adopters of information technologies. In fact, libraries began using technology in the 1930s, when they used punch card sorting machines for circulation and acquisitions. A timeline of the history of library automation can be found in **Table 9.2**.

Early Online Searching

Libraries were among the first institutions to use online searching. Before there was an Internet, there were propriety, stand-alone bibliographical indexing services such as Lockheed's DIALOG and SDC's ORBIT. To a large extent, these

---

[16]In early 2008 the whitehouse.com domain name was used for Republican-slanted news about the forthcoming 2008 presidential election.

databases replaced publications such as *The Reader's Guide to Periodical Literature* and other professional print-based periodical indexes. A library wanting to access a bibliographical database used a modem to dial into the Public Switched Data Network (PSDN), which handled switching to connect with the computers hosting the desired database. During the early years, connections were made using acoustic couplers, modems that required a telephone headset placed into rubber cups (**Figure 9.1**). An acoustic coupler actually "listens" to the tones sent over the telephone line and then to the telephone handset's speaker. To send, an acoustic coupler uses a speaker to transmit the appropriate tones, which are then "heard" by the handset's microphone.[17]

Libraries were charged for database access by the amount of time they spent connected to the service and by the number of citations retrieved and/or printed. Large search results could be printed by the company operating the databases and mailed to the researcher, saving the cost of a long connection. The text of the articles wasn't available online. Instead, searchers received citations, much like those in a bibliography, and in some cases, an abstract.

Unlike the Internet, these early bibliographical databases were not integrated. Each was a stand-alone product, intended to generate income for the company operating it. Librarians chose a database based on the journals it indexed. Sometimes a search required accessing several databases and then integrating the results manually.

Online Circulation Systems

During the 1950s, a circulating library book had a small pocket in the back containing two cards. The author, title, and call number of the book were printed on both cards. To check out the book, a librarian would pull both cards and use an ink pad and dater stamp to stamp the return date on both cards; typically,

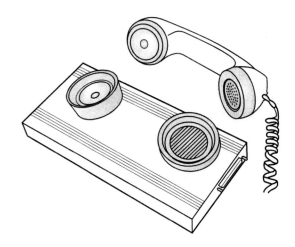

Figure 9.1
An acoustic coupler.

---

[17]Needless to say, an acoustic coupler is prone to a significant amount of transmission errors. Online communications became much more reliable with the development of direct-connect modems that included telephone dialing technology, eliminating the use of a telephone handset.

Table 9.2 Major Events in the History of Library Automation

Date	Event
1930s	Punch card equipment used for circulation and acquisitions.
1961	A computer produces "keyword in context" (KWIC) indexing for Chemical Abstracts.
1960s	Computers produce machine-readable information for cataloging library materials for the Library of Congress.
1965–1968	Library of Congress conducts the MARC I and MARC II projects, which use computers to tag cataloging records by identifying information with three-digit fields.
1967	The Online Computer Library Center (OCLC) begins operation, providing a searchable cataloging database. Initially, it is limited to Ohio.
1968	Demonstrations of DIALOG, an automated bibliographic search tool begin.
1971	OCLC expands to include libraries around the country.
Late 1970s	Independent searchable citation databases, such as DIALOG and ORBIT, become available for widespread use.
Late 1970s	BALLOTS begins operation, providing cataloging information for libraries using the MARC system.
1980	The University of California's online catalog develops into MELVYL, which uses ARPANET (the precursor of the Internet) to make the catalog available nationwide.
1980s	Libraries use minicomputers to run circulation systems.
1980s	Card catalogs begin to go online.
1980s	CD-ROMs first used to replace bibliographical sources formerly in print.

the patron's identifying information was stamped or written from the patron's library card. One of the two book cards went back into the pocket so that you would know when the book was due; the other card was kept at the library on file so that the library knew what was out. The cards retained at the library would be sorted manually by date due and then call number.

When a book was returned, a library clerk found the card for the book and put it back in the book's pocket. The book then could be returned to its place on a shelf. Overdue books were identified by those cards that were still on file after the due date had passed.

Automated circulation systems changed the preceding process considerably. Most commercial systems assign bar codes to circulating items. Each physical item gets a unique bar code, making it easy to track multiple copies of the same item. Patron cards may also have bar codes or some other form of machine-readable identification. Library clerks then can scan the barcodes, just as they would in a retail store. The database behind the circulation system creates logical linkages between the patron and the items being checked out. Because the data are in a database, they can be ordered and searched in a variety ways: by date, by patron, by item, for example. Determining who has a book and when the book is due back becomes a simple task.

Many libraries have established Web sites. Although you do need to make a trip to the library to return the materials you've borrowed, you can often renew loans using the Web site. Most library automation systems also let patrons place reservations for items electronically. When an item is returned, the library automation system produces alerts for items that are overdue or reserved by another patron.

Originally, automated circulation systems used central minicomputers accessed by dumb terminals. Today, however, library systems are either client/server—one or more small servers connected via a local area network (LAN) along with personal computer workstations—or Web based, where access is through a Web browser, which communicates with a Web server and a database.

Automated library circulation systems have only one drawback: When the network or computers go down, checking out items becomes difficult. In that case librarians must write down barcodes on paper so they can be entered into the system when service is restored.

Death of the Card Catalog

If you're younger than 20 years old, you may never have seen a physical library card catalog. Most libraries today—with the exception of some very small libraries—have done away with a physical catalog of book holdings and replaced it with an online database. Card catalogs were usually large wooden cases full of small drawers that held $3'' \times 5''$ cards. The drawers contained an alphabetically organized set of author, title, and subject cards.

Early in the last century, the cards were typed. However, as high-speed computer printers became available, sets of cards could be purchased for most titles (**Figure 9.2**). Regardless of how the cards were produced, they had to be filed by hand, a long and tedious process for large catalogs.

To search a card catalog, you looked in the alphabetic sequence under author, title, or subject. An author card might be located next to subject cards about the author, but a title search could often be isolated from subjects and authors. Today's electronic catalogs provide Boolean searching, where we can AND and OR search terms together, letting us search on multiple terms and, if we so choose, on author, title, and/or subject in the same search.

The online catalog has made searches far more effective than anything that could be done with a physical card catalog. The integration of catalogs and circulation systems has made it possible to let a library patron know immediately whether a book is on the shelf, checked out, and, if checked out, when it's due back.

Figure 9.2
Old-fashioned library
catalog cards.

Text Libraries on the Internet

Before the use of the Web by libraries, a search for a periodical article went like this: Find the paper periodical index and look up articles and then go to the shelves to find the articles you want (or ask a clerk to get the volume for you if the periodical stacks aren't open to the public). Online citation databases changed only the first part of the process by making is easier and more efficient to search for articles: You still needed to find the physical periodical to gain access to the article.

The advent of the Web and the acceptance of PDF as a universal document format standard has begun to change how we access periodicals and, in some cases, entire books. Although the acceptance of electronic books for pleasure reading has been very slow, it has been much greater for nonfiction materials. This is particularly true for professional research publications. For example, when *Chemical Abstracts*, a periodical that indexes publications about chemistry research, was in print, a year would occupy an entire floor-to-ceiling bookcase. The actual journals represented by the index occupied several more bookcases. Providing chemistry research publications, such as might be done in a college or university, was costly in terms of purchasing the index and the journals and also in terms of shelf space.[18]

The first solution to the problem was to provide the articles on microfilm or microfiche. Both require large bulky readers that can be used only at the library owning the film or fiche. In most cases, a copy of an article can be printed for a per-page fee. These microforms solve the problem of shelf space—and concomitantly the problem of pages torn out of journals—but don't provide particularly convenient access. Libraries that have microform holdings continue to use them today but rarely add to them.

Today, indexing services such as *Chemical Abstracts* are totally electronic, providing a variety of databases for researchers.[19] But perhaps more importantly, the full-text of the articles themselves is available over the Web. Most articles have been saved as PDF files. Notice the HTML and PDF links that appear at the left of the window. These provide access to the text of the item in the stated format.

---

[18]The other major problem with the print versions of indexes to scientific literature was the time lag between when a journal article appeared and when the index was available. For example, the index to 1951 chemistry publications did not appear until November 1952. Many of the items abstracted were actually published in 1950, which meant that it took nearly two years for them to be listed in a periodical index.

[19]For more information, see http://www.cas.org/index.html

Full-text research articles are typically not free. Libraries pay a fee for access to the databases. Because of copyright issues and cost, full-text databases usually are not available freely to the public. Users must either be affiliated with the organization purchasing access to the database (typically either college/university libraries or corporate libraries) or must pay a fee for each page or article printed (more typical of public libraries). Nonetheless, electronic access to the full text of periodicals has several advantages over print (and even microform): It costs less than purchasing print journals, it takes no library floor space, and users can do their research from remote locations, usually over the Internet. Electronic periodicals never go missing and they can't be defaced or have pages torn from them; many users can access the same issue at the same time.

The full text of books can be found electronically as well. Libraries pay for access to the databases of the books, which can be searched by authorized library patrons. To retain some control over the use of copyrighted materials, full-text book services typically do not support the downloading of the text. Instead, they provide their own Web-based interface for viewing and copying short passages of text. Most books in electronic full-text libraries are nonfiction, although many fiction titles that are no longer copyrighted (for example, the works of Charles Dickens) are also available. If a book that you want is available in an electronic library to which you have access and you don't mind reading from the computer screen, then this is an excellent way to obtain material. It can't be stolen from the library, can't be checked out, and can't be damaged; it's always there when you want it and you don't have to go physically to the library to get it. The same electronic publication can also be made available to many readers at the same time. The two limitations to such collections are that not all titles are available in this format and many people do not want to read online.

Library Networks

Today's libraries—both public and academic—are also extensive users of wide area networks (WANs). Many regions of the country are linked by computer networks into large systems that can share resources among patrons of any member library. Patrons have become accustomed to the following services:

- Access to a searchable, online catalog that includes holdings from all member libraries. Commonly, this type of integrated catalog is provided over the Web.
- Ability to request any item held by any library in the system, either by phone or online. Typically, library systems of this type send trucks regularly between member libraries to move materials around.
- Ability to renew materials online.
- Ability to return materials to any member library. (The trucks make this possible.)
- Ability to request interlibrary loan of materials from sources outside the home library network.

Unless a person wants to browse, there is rarely any need for a computer-equipped patron to visit the physical library except to pick up and return books or to search using a paper-based periodicals index.

WHERE WE'VE BEEN

The use of technology is a given in Western society. However, there is significant controversy of the role of that technology in children's lives. A group of physicians believes that no child under the age of two should watch television; at the same time, there are television networks and computer software for infants.

Most children have used computer technology by age three. Computers are available at many preschools and are ubiquitous at elementary, middle, and high schools. Educational software is available for children in preschool and elementary school. Older students use the Internet for research and use productivity software for their schoolwork.

The immense amount of research material available on the Internet has made plagiarism a major problem. Web sites that detect plagiarism of Web resources have appeared to help teachers identify copied material.

Libraries have embraced information technology wholeheartedly. Card catalogs in all but the smallest libraries have been replaced with searchable, online databases; periodical indexes and the full text of periodicals and books are often available in electronic form. Libraries also use computers to manage their circulation and join in library networks to facilitate the sharing of resources.

THINGS TO THINK ABOUT

1. Assume you have a child under two years of age. How will you decide what technology you will allow that child to use? Whose opinion do you believe? Why?

2. You are the parent of a six-month-old infant. Your aunt has given you a subscription to a television channel designed for viewing by children less than 18 months old. Will you use this gift? Support your position with the opinions of experts that you locate using Web resources. (Hint: A good place to start this research is www.eric.ed.gov, which has a large searchable database of papers about education.)

3. The U.S. government has declared obesity in children to be a major national concern. One of the supposed culprits is technology use, based on the idea that the more time children spend with technology the less time they spend in physical activity. At least one study (discussed in this chapter) has found the reverse to be true. Conduct research for additional information about the relationship between technology use and physical activity. Does your research support the original study, or does it support the opinion of the U.S. government?

4. Given that so many children use the Internet for recreation and research, should the Internet continue to be unregulated, or should we try to make it safer for children by restricting what material can be posted to Web sites? Why?

5. Assume that you are still in high school and that you have a younger sibling in sixth grade. Your parents are considering installing Internet-filtering software on the family computer (the only one in the house). Create a two-column table that lists the pros and cons of doing so. Based on the comparison in your chart, should your parents go ahead and install the filtering software? Would your answer change if primary school children had access to the computer? Why?

6. When was the last time you went to the library? What did you do? Ask the same questions of someone who is a generation older than you (for example, your parents) and of someone who is two generations older (for example, your grandparents).

7. Think about the last research paper you prepared for school. Where did you start your research? What research media did you use the most? Ask the same questions of someone who is a generation older than you (for example, your parents) and of someone who is two generations older (for example, your grandparents).

WHERE TO GO FROM HERE

American Academy of Child & Adolescent Psychology. "Children online." *http://aacap.org/page.ww?name=Children+Online&section=Facts+for+Families*

Cesarone, Bernard. "Computers in elementary and early childhood education." *http://findarticles.com/p/articles/mi_qa3614/is_200010/ai_n8904161*

Gilbert, Alorie, and Stefani Olsen. "Do Web filters protect your child." *http://www.news.com/Do-Web-filters-protect-your-child/2100–1032_3–6030200.html*

"Guide to First Class Learning Software." *http://www.learningvillage.com/*

"Internet filter review." *http://internet-filter-review.toptenreviews.com/*

Iowa State University e-Library "Student plagiarism websites." *http://lib.iastate.edu/commons/resources/facultyguides/plagiarism/websites.html*

Kafai, Yasmin B., Althea Scott Nixon, and Bruce Burnam. "Digital Dilemmas: How Elementary Preservice Teachers Reason about Students' Appropriate Computer and Internet Use." *Journal of Technology and Teacher Education* 15 (2007): 409–424.

Light, Paul. *Social Processes in Children's Learning.* Cambridge, UK: Cambridge University Press, 2000.

Northwest Regional Educational Laboratory. "Technology in early childhood education: finding the balance." *http://www.nwrel.org/request/june01/child.html*

Schaffer, David Williamson. *How Computer Games Help Children Learn*. Basingstoke, Hampshire, UK: Palgrave Macmillan, 2006.

Shields, Robert. "How do you choose educational software?" *http://www.learningvillage.com/html/artshields.html*

Stewart, Christine. "Computers could disable children." *http://news.bbc.co.uk/2/hi/health/1041677.stm*

Stoerger, Sharon. "Plagiarism." *http://www.web-miner.com/plagiarism*

Science and Medicine 10

WHERE WE'RE GOING

In this chapter we will

- Discover the use of technology as aids to medical diagnosis.
- Look at technologies used in surgery, such as robotic surgery.
- Discuss how technology enhances the functions of the clinical laboratory.
- Learn about the use of technology in many fields of scientific research.
- Talk about medical and scientific technologies that are ethically and religiously controversial.

INTRODUCTION

Technology has always had a strong partnership with scientific and medical research. Much of what has been accomplished in medical research would not be possible without equipment such as the electron microscope, for example. In this chapter we will look at how technology has supported developments in pure scientific research and in medical research. We will also consider a variety of technologies that have been considered controversial.

HEALTH CARE

Technology has had a place in health care for more than 100 years. Whether it improved health care in the first half of the 20th century is open to question, because many of the gains in the survival rate of surgical patients during that period can be attributed to higher standards of cleanliness rather than to specific technologies. Today, however, we rely on increasingly sophisticated diagnostic and surgical technologies to provide solutions for physical conditions that were once thought incurable.

Diagnostic Imaging Technologies

One fairly obvious way a physician can improve care is to ensure that he or she is treating the correct illness or condition. This simple fact has lead to a series of imaging technologies for diagnosis.

The first piece of diagnostic technology was the x-ray machine. The path that led to its development began in 1785, when William Morgan demonstrated x-rays experimentally. Along the way, researchers developed cathode ray tubes—the major electronic parts of the first computers—to produce the rays.[1] Wilhelm Röntgen is credited with identifying the medical properties of x-rays and with building a working machine in 1895. A year later, Thomas Edison produced the fluoroscope, which formed the basis for medical x-ray machines.

X-rays give physicians a two-dimensional view of bones and soft tissues, such as the lungs or brain (**Figure 10.1**). The images allow physicians to identify broken bones, some soft tissues tumors, and foreign bodies that may have been ingested or shot into a person. However, over time we have come to recognize that x-rays carry risks: Overexposure to the radiation produced by an x-ray machine can cause cancer, infertility, and/or birth defects. Although x-rays are still a valuable diagnostic tool today, they are generally used only when necessary, and, in some cases, great care is taken to avoid subjecting the reproductive organs unnecessarily and to shield x-ray technicians from repeated exposures. This is why they almost always stand behind a protective partition when the x-ray machine is turned on.

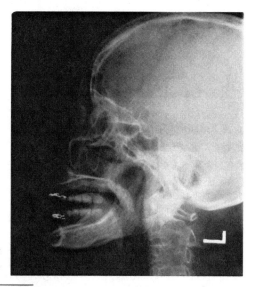

Figure 10.1
An x-ray image
of a skull.

---

[1] The florescence created when streams of electrons hit the glass sides of a vacuum tube was initially thought to be caused by "cathode rays." However, today we know that what experimenters were seeing was really produced by a stream of electrons.

The drawback to x-rays as a diagnostic tool is that they produce a two-dimensional image in which all parts of the body appear to be laying in the same plane. Medical diagnosis could be more precise if an image of a cross-section of the body were available. Computed tomography (CT), once referred to as computed axial tomography, provides such x-ray images. A commercially usable implementation of the technology was developed in England in the mid-1960s by Sir Godfrey Newbold Hounsfield. The device became available publicly in the early 1970s.[2]

A CT image, such as that in **Figure 10.2**, looks like a cross-section created by extracting a thin slice through a body.[3] Although each image itself is still a two-dimensional representation, software today can assemble multiple images into a three-dimensional display.

To take a CT image, a patient is placed on a stretcher and rolled into the scanner (**Figure 10.3**). The patient must keep the part of the body being scanned absolutely still, often requiring physical restraints to prevent movement. For those who suffer from claustrophobia, obtaining an image can be a terrifying experience. Because a CT machine uses x-rays to produce images, patients are also subject to the problems of radiation exposure if the device is overused.

To get away from the use of x-rays, physicians can turn to magnetic resonance imaging (MRI). Like CT, an MRI produces cross-section images (**Figure 10.4**). It uses a magnetic field that magnetizes hydrogen atoms in the body. Radio waves then disrupt the magnetization, which causes the hydrogen atoms to emit a radio signal of their own. The scanner amplifies the waves and uses them to construct an image of the body part being scanned. Software can turn the two-dimensional cross-sections into three-dimensional reconstructions.

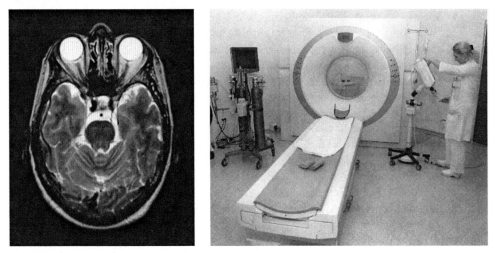

Figure 10.2 A CT image.   **Figure 10.3** A CT machine.

[2] Interesting bit of historical trivia: The company that produced the first CT scanners for sale was EMI, the same company that distributed Beatles recordings. In fact, it was the profit from the record sales that allowed the company to invest in the development of the new medical technology.

[3] CT scanners are used in fields other than medicine. For example, archeologists use them to view the inside of mummies and other artifacts that are too fragile (or precious) to take apart.

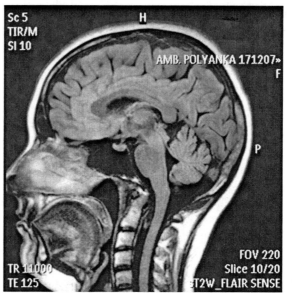

Sc 5
TIR/M
SI 10

H

AMB. POLYANKA 171207»
F

P

FOV 220
Slice 10/20
T2W_FLAIR SENSE

TR 11000
TE 125

Figure 10.4
An MRI image.

Originally, MRI machines looked very like CT machines, a closed chamber requiring that a patient be placed inside and held absolutely still, sometimes for as long as an hour. Today, however, some diagnostic centers have installed "open" MRI machines that do not totally enclose the patient, making it easier for those who are bothered by tight spaces. An MRI also has the advantage of avoiding radiation exposure. Unfortunately, the images generated by the open machines are not as good as those produced by closed machines. In addition, an MRI machine is loud: Patients hear a repeated loud pounding throughout the scan. MRIs may be impossible for some patients without sedation or general anesthetic.

The MRI is best suited to scans of noncalcified tissues, whereas CTs provide superior images of calcified tissues. MRIs are also more expensive and less widely available than CTs. The MRI can also be used repeatedly on the same patient over a short period of time because it does not use radiation.

Sound can also be used to generate images of what is inside a living body. Probably the best known use of ultrasound techniques is for scanning unborn children. The sonogram machine is a small, stand-alone unit (**Figure 10.5**) to which is attached a probe. A technician runs the probe over the area to be scanned. The patient neither needs to lie perfectly still nor be enclosed in any way.

Most ultrasound machines produce two-dimensional black and white images; three-dimensional color images are also available, but because they are more costly, they are not used as frequently. The probe sends sound into the tissues being scanned. The tissues return an echo (much like objects hit by radar beams), which is then interpreted by the machine as an image. Sonograms have made it possible to verify that a fetus is developing properly. It can also be used to identify skeletal abnormalities early in a pregnancy, giving the parents a chance to decide

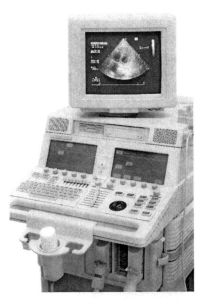

Figure 10.5
An ultrasound machine.

whether to continue the pregnancy. Ultrasound is also used outside obstetrics for diagnosing problems such as tumors and blood clots.

Ultrasounds are generally considered to be safe for developing fetuses. However, some concern has been raised because today pregnant women often have multiple ultrasounds, some of which take place during the first trimester.[4]

In Western societies, doctors recommend the use of specific diagnostic technologies. It is left to the patient, however, to make the final decision as to whether the risks of the technology outweigh its benefits. In some cases—where a broken bone is suspected, for example—the decision in favor of a diagnostic x-ray is relatively straightforward. However, the number of ultrasounds that are safe for a risky pregnancy can be difficult to determine.

Surgical Technologies

Remember the scene in "The Empire Strikes Back" in which Luke Skywalker is treated for injuries he received in the snow? He was healed floating in a tank, attended by a robot doctor. Robot-administered medical procedures have long been a staple of science fiction. To some degree, such surgical robots are a reality, although they require a doctor's hands at the controls.

The current state-of-the-art is exemplified by the da Vinci® Surgical System (a product of Intuitive Surgical®).[5] The system, shown in **Figure 10.6**, consists of three parts: a cart containing the computer and other controlling electronics (the right image in Figure 10.6), the surgeon's console (left), and the surgical arms (center). One of the surgical arms holds a camera that sends high-resolution

[4]See http://www.plus-size-pregnancy.org/Prenatal%20Testing/prenataltest-ultrasoundsafety.htm for more information.
[5]http://www.intuitivesurgical.com/index.aspx

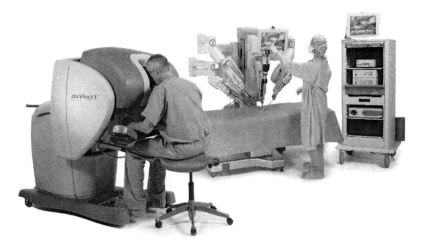

Figure 10.6   The da Vinci Surgical System.

three-dimensional video of the surgical field to the surgeon's console. The remaining arms hold surgical instruments that can be moved with a fuller range of motion than manual laparoscopic tools, which move in only two dimensions.

The da Vinci Surgical System cannot make decisions or move on its own. Sitting at the console, the surgeon views the three-dimensional video of the surgical field and uses the device shown in **Figure 10.7** to direct the movements of the other robotic arms. The robot can perform more precise movements than a human hand, but because it is controlled by a physician, it is considered a surgical tool rather than a robot that performs surgery under human control.

Robotic surgical systems let doctors use minimally invasive surgery in more circumstances than would be possible with traditional instruments. For some patients, this can mean less pain and a shorter recovery time. Because the physician is in control of the robot, patients are typically comfortable with use of such technology. If the technology were to fail, the physician could complete the surgery using manual techniques.

Most robotic surgery is performed with the physician and the robots in the same room. However, there are some instances in which the doctor is hundreds of miles away from the robot.[6] Distance surgery is used in two circumstances: The patient lives in an area too remote for travel to a hospital where the surgical procedure can be performed (for example, rural Alaska or perhaps some day on Mars) and/or the surgery requires a physician with specialized training who is unable to travel to where the patient is located. In the first case, physician assistants and/or nurses typically are present during the surgery, but there is no physician on-site. This presents a risk when the surgery doesn't go as planned. However, when a specialist is performing a distance procedure, the patient is in a hospital operating room and a second physician is physically present to take

---

[6]For more information, see http://www.temple.edu/ispr/examples/ex00_12_28.html

Figure 10.7
A hand control for
the robotic surgical
instruments in the
da Vinci Surgical System.

action if there are problems. The safeguard raises the cost of the surgery because an additional doctor is required. Another problem with distance surgery is the time lag between the video sent from the surgical site to the controlling physician and the return of signals to the robot. Such a delay could be a serious issue for operations performed at great distances, such as in space.

Technology in the Clinical Laboratory

One of the most technology-rich medical subfields is the clinical laboratory. Most of the lab tests ordered by physicians are now processed, at least in part, by machine. Some tests, such as microbiological cultures, must be read by human technicians, although part of the preparation of the test materials may be automated. In contrast, most blood testing is performed by machine. A clinical laboratory today is made up of many computer-driven workstations. Blood analysis (including hematology, chemistry, and immunoassay testing) has been a staple of the medical laboratory for more than 30 years. Developed initially by the Coulter Company (now known as Beckman-Coulter), the machines that perform hematology analysis are still often called "Coulter counters." They provide complete blood counts, differentials, leukocyte counts, and reticulocyte counts. Once blood has been collected, a technician centrifuges the blood tube and then inserts a sample into the machine. The analysis takes about 10 minutes less than it would if done by hand and, in the long run, is more accurate. The results are displayed both on a screen and on paper.

Some medical tests are performed by examining a slide under a microscope. Although the slide is read by a human, the slides are often prepared with the assistance of technology. For example, when a slide needs to be made from an extremely thin piece of tissue, the sample is frozen and then a machine slices off a piece much thinner than could be produced by a human. Many blood analysis machines can prepare "smears" of blood for human reading as well.

The state-of-the-art in clinical laboratory technology currently is a "work-cell," a device that automates a significant range of blood analyses, including standard hematology, blood chemistry, and immunoassay tests. Such equipment accepts tubes of blood and then can perform multiple tests at the same time on the same sample. For high-volume labs, this can save significant amounts of time. The equipment extracts blood through a blood collection tube's plastic stopper, eliminating much of the need for a technician to handle blood.

As noted earlier, medical laboratories use centrifuges and autoclaves (sterilizers). They also need refrigerators, freezers, and incubation equipment that maintain the specific environments for the growth or prevention of growth of lab specimens and store chemicals that must be kept at a specific temperature.

General Purpose Computing in Medicine

The medical technology we have discussed to this point is primarily equipment specifically designed for medical use. However, there is also a significant use of general purpose computing—laptops, desktop, and handheld machines—in medical settings. A small study from 2003 conducted in Britain discovered that physicians and nurses used a handheld computer in the hospital with some enthusiasm.[7] Not only were they able to store protocols for treating specific illnesses, but they found significant utility in more common applications such as a scheduler.

Also consider that hospitals and medical practices are businesses. They have the same needs as any service business for keeping track of customers (patients), appointments, billing, and receipts. In essence, they are selling a service—health care—and therefore tend to use computing for many typical business functions.

The one area in which medical practices seem to be slow in automating is keeping patient records. The penetration of electronic medical records in the United States is very small. In 2006, only 16% of American primary care physicians had implemented electronic patient records systems.[8] Medical records are updated on paper; prescriptions are written on smaller pieces of paper. It's not unusual to go into a medical office and see computers in use for appointment tracking and personal and insurance billing yet patient records are still in paper folders. There are several reasons for this. Although electronic patient records would ease the sharing of information between hospitals and medical offices, there is concern about the privacy of the information.[9,10]

The amount of data currently stored on paper is also a concern: The cost of the conversion of paper records to electronic records is significant. Hospitals and large practices may find it too costly to convert existing records and may began using an electronic system for new records only, creating the possibility

---

[7]For the complete study, see http://www.mo.md/id155.htm
[8]Johnston, Doughlas, et al. *The Value of Computerized Provider Order Entry in Ambulatory Settings: Executive Preview.* Wellesley, MA: Center for Information Technology Leadership, 2003.
[9]For example, see http://www.apapractice.org/apo/pracorg/legislative/the_health_information.html#
[10]As an individual, you can create and store a version of your own medical records online at https://secure9.mymedicalrecords.com/StaticContent/know.html?cpo=2. These data can then be downloaded to a medical practitioner. However, because the data are stored on someone else's computer, privacy is a significant concern.

that existing patients will have records in two places. This means that staff must check for a paper record (and hope they don't overlook it) as well as an electronic record. Along with the cost of conversion, medical offices must also deal with the resistance of medical workers to the change from paper to electronic records.

Adding to the preceding issues is that there are many software vendors that sell electronic medical records systems. All these products are proprietary, and there is no guarantee that the formats in which they store data will be compatible, making it difficult to exchange data between different systems. Unlike file formats such as PDF, there is no standardization.

Government regulations have also not caught up with medical records technology. In many countries, including the United States, medical records must be kept in their original form and not be changed. (They can be amended, but not edited.) They must be signed to verify their authenticity. To do this electronically, there must be some way to prevent editing an existing record and an acceptance of a digital signature in lieu of physical handwriting. This last issue alone makes it extremely difficult to keep medical records solely on a computer.

TECHNOLOGY IN SCIENCE

Scientific research today is steeped in technology. It has allowed scientists to view the tiniest elements of our world, to combine dangerous chemicals safely, to explore the cosmos, and to predict dangerous storms on Earth. There is so much technology being used in scientific research that it is impossible to discuss it all. However, this section gives an overview of the types and uses of technology in the pure sciences.

Biology: Looking at Small Things

Because it deals with living organisms, and often very small parts of living organisms, biological research has used technology for centuries. In particular, the first microscope appeared in the 17th century. It was known as a "light microscope" because it used light (first natural and later from an electric light bulb) to illuminate the item being viewed.[11] Early microscopes had only a single lens. Compound microscopes, with multiple lenses, appeared about 150 years later. Compound microscopes are widely used today, with three lenses that typically magnify 4, 10, and 40 times (**Figure 10.8**). Some also have a fourth lens that requires a drop of oil between the lens and what is being viewed (an "immersion lens"). Immersion lenses typically provide 100× magnification.

With a maximum magnification of 100×, a light microscope can't provide enough detail for advanced biological research. Researchers want to look at human chromosomes, for example, rather than being limited to examining entire cells. Electron microscopes (**Figure 10.9**), however, provide magnification of up to 2 million times. One such instrument (a scanning electron microscope) produced the image of an ant shown in **Figure 10.10**.

---

[11] There is some controversy as to who actually invented the microscope. Zacharias Janssen, who claimed to have invented it in 1590, was actually born in 1590, making his claim a bit dodgy. However, Anton von Leeuwenhoek is commonly viewed as the person who introduced the microscope to biologists in the mid-17th century.

Figure 10.8   A compound microscope.   **Figure 10.9**   An electron microscope.

Figure 10.10 An image taken through an electron
microscope.

There are actually four technologies used in electron microscopes:

- Transmission uses an electron beam shot through an item that has parts that are transparent. The spacing of the beams that come through the item forms the image for the user to view.
- Scanning uses an electron beam that scans across an item, much in the way a scanner captures the image on a piece of paper.
- Reflection uses the reflections of an electron beam from an object to create an image.
- Scanning transmission adds scanning to the technology of the transmission electron microscope.

There is little controversy surrounding the use of microscopes in and of themselves in biological research. However, the types of research that microscopes enable have been controversial for centuries. For example, when light microscopes first made it possible to examine the elements of blood (red blood cells, white blood cells, platelets, and so on), some people branded the work as "unnatural." Each time technology has enabled us to drill down to smaller parts of the human body, some segment of the human population has objected. On the positive side—and this is a very big positive—microscopes are fundamental in the processes of identifying and curing illnesses.

Atmospheric Sciences: Predicting the Weather

Most of us are familiar with weather forecasts. Sometimes we see them on television or read them in the newspaper; more frequently today we view them on the Web. The technology behind the weather maps and precipitation percentages is a combination of hardware and software. Much of the software originates with the U.S. National Weather Service,[12] which began using mainframes in the 1950s to compute forecasts using mathematical models.

The input data for the computer models comes from a variety of weather instruments. In addition to thermometers, they include the following instruments:

- *Anemometer:* Measures wind speed. Invented in 1846, by John Thomas Romney Robinson, the first devices caught the wind in cups. Current technologies include
  - Heated wires that measure the temperature drop from the wind
  - Measuring the difference the Doppler shift between a laser inside the device and one outside the device that is affected by the wind
  - Measuring the difference between sound waves emitted from two or three transducers (sonic anemometers)
  - Plates and/or tubes that measure the pressure created by moving air
- *Barometer:* Measures atmospheric pressure. The barometer was invented in the mid-17th century by Gasparo Berti; the first barometer to use mercury appeared in 1643. Current barometers use water, mercury, and cells of a beryllium/copper alloy that expand and contract with changing air pressure (aneroid barometers).

---

[12]See http://www.research.noaa.gov/weather

- *Hygrometer:* Measures humidity. Most hygrometers use two thermometers, one that is kept dry and the other wet. As the water evaporates from the bulb of the wet thermometer, its temperature drops. The difference between the readings on the dry thermometer and the wet thermometer is used to compute relative humidity.
- *Rain/snow gauge:* Measures the amount of precipitation. Simple precipitation gauges are little more than calibrated tubes.
- *Weather radar:* Reports the location, intensity, direction of movement, and type of precipitation. Weather radar stations use the Doppler effect and are often called "Doppler radar." The software that processes the radar information prepares the weather maps with which most of us are familiar.
- *Satellite:* Takes photographs of atmospheric elements such as clouds over time to provide input for mapping movement. Satellites, like that shown in **Figure 10.11**, also can be used to photograph fires (for example, forest fires), dust storms, ocean currents, snow and ice cover, and so on. The individual images are put together to provide the satellite weather maps that we commonly see with weather forecasts.

The data collected from weather instruments are combined to produce weather forecasts. The most sophisticated type of forecast—a numerical weather prediction—requires the power of a supercomputer and data such as temperature, air pressure, wind speed and direction, and type and amount of precipitation. However, the numerical models that produce the forecasts aren't perfect, and even with significant computing power forecasts are often wrong. The farther away in time a forecast, the higher the chance that it will be inaccurate.

What does all of this weather-related technology buy us? The forecasts are essential for a number of industries, including aviation, sailing, and farming. Storm

Figure 10.11
The GOES-1, a U.S. weather satellite.

warnings often make it possible to alert people in time to evacuate areas in the paths of dangerous weather such as hurricanes and tornadoes. However, as many of us have discovered to our chagrin, forecasts aren't necessarily accurate. Most of the time an inaccurate forecast means little more than getting wet because we didn't bring an umbrella or being hot because we dressed too warmly. On the other hand, when forecasts inaccurately predict the severity of a storm, lives can be at risk.

Astronomy: Exploring Space

Astronomy is one of those sciences that didn't get very far without the aid of technology. Beyond what we can see with our naked eyes, standing on earth, everything we know about the cosmos comes from the use of telescopes, computer analysis of the data they provide, and the small distance humans have traveled from Earth.

Initially, studying space was not well accepted because the Catholic Church supported the belief in an earth-centered cosmos. Copernicus, the most vocal early proponent of the sun-centered theory, was labeled a heretic. His observations, however, were made with the naked eye and there was little solid proof of the validity of his theory. Galileo (**Figure 10.12**) invented the refracting telescope in 1609 and was then able to view phenomena such as the movement of the planet Venus to provide scientifically verifiable evidence for the correctness of Copernicus' theory.

In 1616, Galileo was ordered by the Inquisition to stop defending the Copernican theory of the cosmos, despite the support of Cardinal Maffeo Barberini (later Pope Urban VII). The Inquisition declared that it was not possible that the earth moved around the sun; Galileo was not to publish anything about his heliocentric theory without its being cleared by Roman censors. By 1624, however, the Inquisition had relented enough to allow him to publish as long as he put forth his ideas as theories rather than facts. Galileo's 1630 manuscript discussing his research findings didn't receive a full examination by the censorship board: The Bubonic Plague struck and therefore only the preface and the end of the book were actually submitted.

Figure 10.12
Galileo Galilei.

When the Pope was made aware of the contents of the book, he referred it to the Inquisition, which then formally charged Galileo with heresy. Galileo, who at age 68 was not in good health, nonetheless traveled to Rome to defend his work before the Inquisition. The Inquisition would not back down from its position that the idea of a sun-centered solar system and a moving earth were heretical. After being threatened with torture, Galileo publically recanted his support for Copernican theory and spent the rest of his life under house arrest.

During Galileo's life, humans ran a real risk that scientific research and technical development would be halted by religious opposition. As the influence of the Inquisition waned, however, researchers continued to refine telescopes and concomitantly learned more about the organization and movement of the cosmos. Refracting telescopes, those based on a technology from the 17th century, remained state-of-the-art until the 20th century, when radio telescopes were developed to provide even better glimpses of what lies beyond out planet.

Today's telescopes are land- or space-based. The largest land-based telescope is the European Space Observatory's Very Large Telescope (VLT), located in the Chilean desert on a mountain named Cerro Paranal.[13] The VLT (**Figure 10.13**) is actually an array of four telescopes with 8.2-meter lens openings that can be used individually or as a unit. The problem with land-based telescopes, however, is that images are blocked and/or distorted by the Earth's atmosphere.

Space-based telescopes avoid problems with atmospheric distortion. To this date, there has been one very successful orbital telescope: the Hubble telescope (**Figure 10.14**).[14] No other instrument has been able to provide comparable images of everything from our own solar system to distant galaxies. This telescope, along with the software that analyzes the images it sends back to Earth, have given scientists information about the birth of the universe by recording light traveling so far that it originated in time close to the "Big Bang." It has provided evidence for phenomena that previously were only theories. For example, it has confirmed the existence of dark matter (**Figure 10.15**), although scientists still aren't sure what dark matter is.

Figure 10.13
The Very Large Telescope (VLT) located in Chile.

[13]For more information, see http://www.eso.org/paranal/
[14]Details can be found at http://hubble.nasa.gov/

Figure 10.14
The Hubble space telescope fully deployed in orbit.

At first glance, knowledge of the cosmos doesn't really have a major impact on the human condition here on Earth.[15] However, telescopes, satellites, and associated instruments have, for example, allowed us to track the size and location of the hole in the Earth's ozone layer.[16] The presence of the ozone layer is essential to the continuation of life as we know it and therefore monitoring the ozone—and possibly taking action to maintain it—is undeniably important.

Despite all the evidence provided to us by modern telescopes and space flight, there are still those who believe that the Earth is flat. The most visible proponents are members of the Flat Earth Society, whose model of our planet can be seen in **Figure 10.16**.[17] Their arguments include concepts such as the presence of "ether" in space (the idea that space is not a vacuum), the lack of gravity if the Earth circled the sun, the assertion that people would fall off the bottom of a round planet, and so on. Like many other groups that support theories that run contrary to current scientific evidence, they point to a conspiracy to keep the public from knowing the truth. In the face of evidence provided by photographs from space missions, they assert that manned space flight is a hoax. Because the preponderance of evidence supports current scientific theory, most people in Western societies see those who support the idea of a flat Earth as ridiculous.

Physics: Discovering What Things Are Made Of

Although astronomy without technology is a rather primitive science, it can be practiced to some extent without technology. Some fields of research, however, would not exist without the aid of man-made equipment. Particle physics is just such a discipline. It provides information about the interior structure of atoms, including the nature of subatomic particles; particle physicists are also involved in researching the Big Bang (the event that is believed to have generated our universe). The particles studied are extremely small. To gain information, atoms or their contents must be smashed together at high speeds (often close to the speed of light).

---

[15] We humans have always been driven to find out about our universe. "Knowledge for knowledge's sake" is much of the motivation behind a great deal of cosmological research. There doesn't necessarily need to be a practical application of the knowledge for people to simply want to know and understand.

[16] Further information can be found at http://www.ozonelayer.noaa.gov/

[17] See http://theflatearthsociety.org

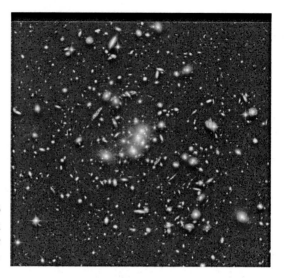

Figure 10.15
An image of dark matter taken by the Hubble telescope.

Figure 10.16
The Flat Earth Society's view of the Earth.

The technology that speeds up the atoms and allows them to collide is known as a "particle accelerator." A collision may produce split atoms for analysis of the interior makeup of an atom or it may produce a new, combined atom. Many of the transuranium elements[18] are produced by allowing smaller atoms to collide and combine.

Because atoms that are to be split or joined must be traveling extremely fast, particle accelerators are generally quite large. For example, in **Figure 10.17** you can find an aerial photo of the Tevatron, a collider located at the U.S.'s Fermi National Accelerator Laboratory.[19] The ring in the photo marks the dimensions of the accelerator—6.3 kilometers. As they travel around the ring, particles can reach speeds of up to 320 kilometers per hour less than the speed of light.

[18]Transuranium elements are elements that have atomic numbers greater than uranium and that don't occur naturally on Earth. All of them are radioactive.
[19]For more details, see http://www-bdnew.fnal.gov/tevatron/

Figure 10.17
An aerial view of the Tevatron accelerator at Fermilab.

The Tevatron is not, however, the largest particle accelerator in the world. That title belongs to the Large Hadron Collider (LHC), constructed by the European Organization for Nuclear Research.[20] The LHC is a 27-kilometer ring. The ring is so large that technicians need to use some kind of vehicle to reach portions needing service. Major construction on the LHC was completed in November 2007, with operations beginning in mid-2008.

Many of the products of the collision of atoms are very short lived. For example, roentgenium, a transuranium element with an atomic number of 111, has a half-life of only 3.6 seconds.[21] Therefore the equipment that produces the particles must include instrumentation to monitor and record the results of a collision. Particle physics is certainly a discipline that could not exist without significant technology.

For the most part, particle physics is what we consider to be "pure" scientific research in that it is not necessarily expected to lead to practical applications. Some may argue that there is no reason to conduct such research, especially considering the expense of constructing and operating the large particle accelerator facilities. On the other hand, many people believe that there is value in all knowledge of our universe and that although the things we discover from such research have no "use," the results give us a deeper understanding of the universe in which we live and for that reason alone have value.

CONTROVERSIAL MEDICAL AND SCIENTIFIC TECHNOLOGIES

No other technologies seem to generate as much ethical and religious controversy as medical and scientific technologies. Technology develops faster than society can adapt to the effects of the technology, opening the door for significant—and often

---

[20]Details can be found at http://public.web.cern.ch/Public/en/Spotlight/SpotlightConnectingLHC-en.html
[21]The half-life of a radioactive element is the amount of time it takes for half of any quantity to decay.

passionate—discussion on the appropriate use of new technologies. In this section we look at a few of the current scientific and medical technologies that have generated significant controversy over their use. In these examples, there is no clear right and wrong; our opinions of these technologies are changing continually as research into these technologies proceeds and practical applications emerge.

Cloning

Cloning is the creation of a living organism from the cell of a single parent. The offspring (plant or animal) is genetically identical to its parent. Cloning has long been an accepted technology as part of science fiction stories, but the reality is much more complicated.

Farmers have been cloning plants for years, producing hardier, more disease- and pest-resistant crops. We generally accept plant cloning without question. However, cloning of animals and humans is a different matter. Science is not at the point where we can clone a human organ, much less an entire human being. However, animals have been cloned successfully.

The first cloned animal, Dolly, a sheep, was born in 1996 at the Roslin Institute in Edinburgh, Scotland.[22] Scientists created her by taking the nucleus from a sheep's egg and replacing it with the nucleus of a cell from Dolly's "clone parent." The egg, which then had a complete set of genetic material rather than the half found in mammal eggs, was implanted in another sheep, which acted as a surrogate mother. Currently, all animal clones are produced in this way.

Dolly seemed to be a normal, healthy sheep in all respects. She had four lambs and died at the age of six from a lung disease. (Most sheep live 11–12 years.) Her death had nothing to do with her being a clone; the lung disease from which she suffered is common among sheep.

Since Dolly, researchers have cloned mice, pigs, dogs, cats, tadpoles, carp, Rhesus monkeys, cattle, and horses. However, cloning is not a particularly efficient technology: Cloning efficiency is generally between zero and 3%.[23] In other words, if scientists implant 100 cloned eggs in surrogate mothers, at most three will result in live offspring. Total failure is a much more likely result.

Although a clone is genetically identical to its parent, it may not look or behave exactly like its parent. For example, the first cloned cat ("CarbonCopy," cloned at Texas A&M University in 2001) is white on her legs, chest, and belly; the top of her face and her back are brown with black tabby stripes. Her clone mother, however, is a calico: She has white on her legs, chest and belly, like CarbonCopy, but the top of her face and her back are covered with areas of orange and brown. The brown does have the black tabby stripes. CarbonCopy's personality is also not identical to her clone mother's. There are several reasons why this might be so:

- Not all genes are used to determine how a mammal looks; there are many genes that aren't used. The appearance of an animal such as a cat depends on which of the genes that regulate coat color, texture,

---

[22]http://www.roslin.ac.uk/
[23]For a summary of cloning efficiency, see http://www.mindfully.org/GE/GE3/Cloning-EfficiencyAug01.htm

and patterns are turned on, or "expressed." The same is true for genes that affect personality.

- Personality is determined, at least in part, by environment as well as by genetics, although the genetic component is now considered to be the stronger influence. Most of the evidence for this conclusion comes from studies of identical twin humans raised apart.[24]
- Mutations occur after the embryo is implanted in the surrogate mother. Such mutations have resulted, for example, in one of a pair of identical twins being healthy and the other suffering from muscular dystrophy.

Cloning conceivably could be used to:

- Create genetically identical food plants and animals that are resistant to pests and disease. Such use occurs today.
- Re-create a beloved pet after the animal has died. Although a cat has been cloned to replace a dead pet, the cost was so high (around $32,000 per clone) that the company that produced the clone went out of business. Even if the cloning process were less expensive and reliable, as was mentioned earlier in this chapter, there is no guarantee that a cloned pet will look or behave exactly like its clone parent.
- Preserve endangered species. A guar (a rare wild ox) was cloned in 2001 using cells from an animal that had been dead for eight years. However, the clone lived only 48 hours. Such a technique may be useful but is currently not very effective.
- Re-create species that have become extinct. The idea of re-creating extinct animals brings to mind the fictional cloned dinosaurs of the "Jurassic Park" movies. However intriguing the possibility, scientists say that the DNA in most fossils is incomplete and/or damaged and therefore cannot be used for cloning.[25]
- Grow and store replacement organs and tissues for humans. For example, a few cells extracted from your heart could be used to grow a replacement heart that would be stored in case you needed a heart transplant; cloned skin could be used as skin grafts to help heal burns. Because the cloned replacement parts are identical to the recipient, there would be virtually no chance of rejection.
- Create complete clones of a human being. This is viewed as one way to let people who could not otherwise be biological parents have children that are genetically related to them.

Cloning has been controversial since before Dolly's birth. Among the reasons for opposing cleaning are:

- Religious objections. Western religions tend to view human cloning as "playing God" because only God can create life. This attitude is not found in many Eastern religions that include the idea of reincarnation,

---

[24]Twin studies are fascinating. For an overview, see http://www.townonline.com/parentsandkids/news/x2058360157

[25]Replacing the missing part of the genetic code with the genes of a modern animal—such as the frog DNA used in the "Jurassic Park" movies—does not appear to be a feasible technique.

such as Hinduism and Buddhism.[26] Much of the human cloning research has therefore migrated to Asia.

- The lack of information about the safety of eating cloned plants and animals. We could be subjecting ourselves to health problems as cloned materials built up in our bodies.
- The low success rate. Currently, cloning of most animals is too expensive to make it a reasonable technique to use regularly.
- Decreases in genetic diversity. One of the characteristics that gives a natural population (be it plant or animal) a survival advantage is genetic diversity. For example, a disease that affects even a large proportion of a population will often leave a few immune individuals alive to carry on the species. If all members of a species are identical, the entire species could be wiped out by the same disease or could perpetuate a genetic problem forever.
- Fast clone aging. Dolly's early death and observations of other cloned animals have caused some researchers to wonder whether clones age faster than noncloned animals.

This planet has no worldwide government.[27] Therefore government regulations allowing and forbidding cloning differ throughout the world. Great Britain, for example, passed legislation in 2001 to allow research into therapeutic cloning, cloning that would lead to cures for disease and injury.[28] Human cloning of embryos is permitted, but only until 14 days of age. Australia enacted a similar law in 2007.[29]

The situation in the United States is a bit more complicated. Although several states have enacted laws banning cloning research, the federal government has not done so directly.[30] It has instead banned the use of federal funds for cloning research, allowing the research to proceed using foundation and private funds.

Although Eastern religions may not be opposed to human cloning, some Asian countries, such as South Korea, have enacted legislation that prohibits human cloning but allows therapeutic cloning. China's government supports therapeutic cloning research as well. Nonetheless, Asian cloning research labs tend to have better funding, better equipment, and fewer restrictions than those in the United States.

One of the fascinating questions that surrounds the legal status of cloning research is how we, as a human race, came to the conclusion that legally we should permit therapeutic cloning but not the cloning of entire humans. No single person stood up and imposed existing laws on us. In fact, this world is so politically polarized that a global consensus of opinion on a controversial issue is virtually impossible. Yet that seems to be what has occurred. Without a planet-wide leader or law-making legislative body, many major countries of this world have come to a similar conclusion with regard to the ethics of cloning.

---

[26]For an editorial on this issue, see http://www.nytimes.com/2007/11/20/science/20tier.html
[27]Although we do have the United Nations, its recommendations/resolutions are not binding on any country: They do not have the force of law. However, in 2005 the UN did pass a resolution that sought bans to human cloning.
[28]See http://www.nytimes.com/2007/11/20/science/20tier.html?_r=1&oref=slogin
[29]See http://www.theage.com.au/news/national/state-move-for-cell-cloning/2007/03/12/1173548107146.html
[30]See http://vienna.usembassy.gov/en/download/pdf/human_clone.pdf

There is, as you might expect, more than one way to view this result. We might congratulate ourselves on being an ethical species that can recognize universal truths when necessary. Alternatively, we might see the people of the world as the proverbial sheep, being unable to form our own opinions but instead following the lead of other people and nations. As with many questions that involve morality and ethics, there are as many correct answers as there are religions and political viewpoints in the world.

Stem Cell Research

Closely related to cloning is the science of stem cells, the cells that in a developing embryo change into all the types of cells found in the body. Stem cell therapies have been found to be effective in treating conditions such as Parkinson's disease; the use of stem cells is also being investigated as a potential cure for type 1 diabetes. The hope is that the development of stem cells into needed tissues can be stimulated outside the human body. Such tissues could be used to grow skin for burn victims or to grow replacement limbs or organs, as well as replacing damaged nerve tissues for spinal cord injury victims.

There are two main sources of stem cells: embryos and adult bodies. Although researchers are discovering that adult stem cells occur in more tissues than previously thought, they are small in number and more difficult to force to differentiate to a desired form. The richest source for stem cells, however, is embryos young enough that the cells have not started to differentiate.[31]

There seems to be little controversy over the use of adult stem cells, but most of the research to date has involved embryonic stem cells. Because embryos are destroyed when stem cells are extracted, the process has generated opposition from those who believe that embryos even a few days old are a form of human life.[32] Many embryonic stem cells are produced by cloning from existing stem cell lines. The controversy centers on both the cloning aspects and the creation of new lines directly from embryos.

There are two major ethical questions surrounding the creation of new stem cell lines. First, can new lines be created from frozen embryos that will not be implanted in a woman? The answers to this are closely tied with debate over the disposal of embryos that won't be used. Depending on religious and ethical viewpoints, individuals may accept or reject donating unused embryos to an infertile couple, donating the embryos to scientific research (including stem cell research), or straight-forward destruction of the embryos.

Second, can embryos be created simply for purposes of research and/or medical treatment? Those who oppose the use of unused embryos created for infertility treatments almost uniformly oppose creating embryos specifically for stem cell research. In addition, many of those who support using unwanted embryos in research still oppose creating embryos solely for that purpose.

Public policy with regard to stem cell research varies throughout the world. The United States, for example, does not specifically bar embryonic stem cell

---

[31]The optimum time to harvest human stem cells seems to be four to five days after fertilization.
[32]For a nonhysterical look at opposition to embryonic stem cell research, see http://www.stemcellresearch.org/

research. However, in 2001 then U.S. President George W. Bush enacted a policy that restricted federal research funds to projects that used existing stem cell lines (those created before the enactment of the policy). A summary of the positions of many countries on stem cell research appears in **Table 10.1**.

Reproductive Technologies

Technology to assist in conception and gestation—specifically artificial insemination, in vitro fertilization, and the use of surrogate mothers—has long been controversial. Technologies that prevent conception or that abort a conception are even more controversial. Most of the objections are religious but have effects beyond personal beliefs and actions.

Table 10.1 Positions of Major Nations Regarding Stem Cell Research

Category	Definition	Countries
Permissive	Little or no restrictions on stem cell research	Australia Belgium China Israel Japan Singapore South Korea Sweden United Kingdom
Flexible	Research permitted using only donated embryos from fertility clinics	Brazil Canada France Iran South Africa Spain The Netherlands Taiwan
Restrictive	A continuum of restrictions ranging from a complete ban on stem cell research to limitations to existing stem cell lines	Most restrictive: Austria Ireland Norway Poland Less restrictive: Germany Italy United States

Data from http://www.mbbnet.umn.edu/scmap.html

Assisted Conception The heartbreak of those who wish to mother or father children and are unable to do so has long been a part of the human condition. In fact, the first recorded artificial insemination of a woman with donor sperm occurred in 1884.[33] It certainly isn't a new technology. In vitro fertilization began in 1977 with the conception of Louise Brown.[34] Today, success using either technique is far from certain. The average success rate for artificial insemination is between 5% and 20%, depending on the age of the woman and whether she takes fertility drugs before insemination.[35] The success rate for in vitro fertilization of women under 35 is 35%, but that drops to less than 10% for women over age 40.[36] Both types of assisted conception are relatively expensive. Artificial insemination can cost up to $4,000 per attempt, and in vitro fertilization can cost as much as $12,000 per attempt. Both techniques are therefore limited to those with the means to pay for them or whose insurance covers assisted reproduction.

Because assisted conception is expensive, it is not available to most people living at low economic levels. This type of discrimination by income is not limited to reproductive technologies, however; most technologies, especially when new, are expensive. As mentioned earlier, the largest objection is religious and comes primarily—or at least, most vocally—from the Catholic Church and the Southern Baptists.[37] Their objections include the following reasoning:

- A human embryo is a person at conception and therefore has all the rights of a person.
- Human sexual relations have two functions: unitive (uniting a man and woman and their family) and procreative (Catholic doctrine only).
- In vitro fertilization violates the unitive purpose of sexual relations and deprives a child of the "filial relationship with his parental origins and can hinder the maturing of his personality. It objectively deprives conjugal fruitfulness of its unity and integrity, it brings about and manifests a rupture between genetic parenthood, gestational parenthood, and responsibility for upbringing. This threat to the unity and stability of the family is a source of dissension, disorder, and injustice in the whole of social life."[38] (Catholic doctrine only.)
- The destruction of unimplanted embryos is murder because although they may be only a few cells, they still constitute a human life.

The preceding applies to in vitro fertilization using the sperm and egg of the parents as well as artificial insemination and the use of surrogate mothers. The Catholic Church has further objections to the technique when it involves donor sperm or eggs and/or embryos used in scientific research.

---

[33] For details see http://fubini.swarthmore.edu/~WS30/WS30F1998/nrosad03.html
[34] You can find a history of in vitro fertilization at http://fubini.swarthmore.edu/~WS30/WS30F1998/nrosad03.html
[35] http://www.babycenter.com/0_fertility-treatment-artificial-insemination-iui_4092.bc
[36] http://health.yahoo.com/reproductive-treatment/in-vitro-fertilization-for-infertility/healthwise-hw227379.html
[37] For the Southern Baptist Convention's position on a variety of controversial medical technologies, see http://www.johnstonsarchive.net/baptist/sbcabres.html
[38] http://catholicinsight.com/online/church/vatican/article_475.shtml

Contraception There are two aspects of contraception that can be controversial. The first is the morality of the use of contraception to limit the number of children. The second is its use as a population control method and whether such contraception should be imposed by a government.

Several religious groups object to contraception in some way, based on the Biblical injunction to "be fruitful and multiply." Objectors include the Catholic Church[39] and Orthodox Judaism.[40] The Catholic Church and the Mormon Church forbid all contraception except abstinence.[41] Orthodox Judaism forbids contraception that uses physical barriers; chemical contraception (for example, birth control pills) is acceptable. This attitude is shared by some Christians, including the Jehovah's Witnesses. Islam, on the other hand, does not forbid any method of contraception.[42] Christians such as the Methodists and Lutherans also accept all forms of contraception, although Lutherans object to contraception being used to keep a couple childless.[43]

Buddhism, whose followers believe that it is wrong to kill regardless, permits contraception when it prevents conception (for example, birth control pills or condoms) but opposes it when it destroys an embryo (for example, an interuterine device or the "morning-after" pill).[44] Hinduism (and Sikhism) allows most forms of contraception.[45]

Politically, virtually all countries in the world allow contraception. However, the availability of effective contraception is not spread equally around the globe. The developing world, where population pressures are more acute than in developed nations, are those whose citizens are less likely to have access to family planning of any kind.

In most of the world, contraception is voluntary. China and India, the two most populous nations with over one billion citizens each, are seriously concerned about overpopulation. They have different approaches to population control, which is primarily an effect of the type of government. India is a democracy and, as such, will not pass legislation requiring contraception or limiting the size of families. Instead, India has tried public education campaigns and paying men to have vasectomies.[46] India has also legalized "emergency contraception" methods, such as the morning-after pill.[47] However, because India has so much poverty, contraception often does not reach the neediest. Therefore, despite recent economic gains, population pressures continue to remain a significant problem.

China, on the other hand, has a totalitarian government that can impose contraception requirements as desired. The result has been the "one child per family" rule, in which most families are limited to a single child.[48] There are

---

[39] http://www.lisashea.com/lisabase/aboutme/birthcontrol.html
[40] http://judaism.about.com/od/sexinjudaism/a/birthcontrol.htm
[41] http://www.geocities.com/swickersc/mormonfam.html
[42] http://www.crescentlife.com/family%20matters/islam_and_abortion.htm
[43] http://www.lcms.org/pages/internal.asp?NavID=2122
[44] http://www.bbc.co.uk/religion/religions/buddhism/buddhistethics/contraception.shtml
[45] http://www.blurtit.com/q889025.html
[46] http://www.dancewithshadows.com/society/understanding-contraception.asp
[47] http://www.kaisernetwork.org/daily_reports/rep_index.cfm?DR_ID=32376
[48] http://www.overpopulation.com/faq/countries-of-the-world/asia/china/chinas-one-child-policy/

some exceptions: Farm families, in particular, may have a second child when the first child is a girl. Families that have a second child in defiance of the edict are assessed a heavy fine, something affordable only by the very wealthiest Chinese.

The one-child policy has curtailed China's population growth significantly. It is not, however, without some serious potential problems. Culturally, China has long had a preference for male children. The one child policy has therefore resulted in the death of many female infants and the surrendering of others for adoption. The current generation of Chinese is therefore heavily male. There is concern that there will not be enough women for these men to marry and that, with the one child policy, China's population may drop significantly after the next generation. Of less importance but also worrisome to Chinese citizens is that each single child will be responsible for caring for two aging parents, without the help of siblings.

In some parts of the world, population pressures are exactly the opposite: The birth rate is not high enough to keep up with the death rate and populations are falling. Russia in particular is seeing a drop of nearly one million people a year.[49,50] During the years of the Soviet Union, contraception was not widely available in Russia. Women therefore resorted to abortion as a method of contraception.[51] However, as contraception has become more available, the abortion rate has declined. Nonetheless, educated women are still limiting family sizes and the population continues to fall. It has been suggested that Russians follow a number of strategies: give tax breaks for large families, increase immigration, and pass legislation that is intended to cut the death rate (for example, requiring the use of seatbelts).[52] The Russian government has been resistant to most suggestions, especially that of increasing immigration.

Abortion Of all the reproductive technologies that people consider immoral, abortion seems to engender the greatest passion, evinced by the bombing of clinics where abortions are performed and the murder of doctors who perform abortions. Most major religions have taken public positions on the issue. A summary can be found in **Table 10.2**.

Because abortion is so controversial, it has become important in the political and religious arenas. The more influence a religion has in a country, the more likely it is that the government will implement laws that conform to the position taken by that religion. For example, Italy, the center of the Catholic Church, was among the last of the European countries to allow abortion.[53] Abortion continues to be illegal in Ireland (both the Republic of Ireland and Northern Ireland), which is also heavily Catholic.

---

[49]http://www.csmonitor.com/2002/0418/p06s02-woeu.html

[50]Other former Soviet republics—including Kazakhstan, Ukraine, Belarus, Moldova, Estonia, Latvia, and Georgia—are facing similar issues. Some Eastern European countries (Bulgaria, Bosnia, Croatia, Slovenia, Hungary, Lithuania) also suffer from declining populations. The African nations of Swaziland and Botswana have population declines caused by deaths from AIDS.

[51]http://www.rand.org/pubs/research_briefs/RB5055/index1.html

[52]http://news.bbc.co.uk/2/hi/europe/4125072.stm

[53]http://www.ben.iss.it/precedenti/aprile/1apr_en.htm

Table 10.2 Positions on Abortion of Major World Religions

Church	Position on Abortion
Catholicism	All abortion is forbidden, including abortion to save the life of the mother.[*]
Southern Baptist	All abortion is forbidden, except to save the life of the mother.
Methodist	Abortion should be avoided whenever possible. However, it is acceptable when the fetus is seriously deformed, the pregnancy was the result of rape, or the mother's health is at risk.[†]
Lutheran	Abortion is forbidden, except to save the life of the mother.[‡]
Mormonism	Abortion is forbidden except in cases of rape or incest, where the fetus has a deformity that will make it nonviable at birth, or where the health of the mother is at risk.
Judaism	Judaism views human life as beginning at birth and therefore permits abortion where not doing so would be more detrimental than terminating the pregnancy.[§] Abortion to save the life of the mother or when the fetus would not be viable, even at term, is permitted; abortions for genetic abnormalities that would not cause death at or shortly after birth, such as Down's syndrome, are not. The more Orthodox the Judaism, the less abortion is tolerated.
Islam	In general, all abortion is forbidden, except to save the life of the mother. However, some scholars permit abortions in the first 40 or 120 days of gestation if a deformity would lead to nonviability or if the child would be so disabled that its care would be a significant burden on its family.
Hinduism	All abortion is forbidden, except to save the life of the mother.[¶]
Buddhism	Although Buddhism generally disapproves of abortion, scholars state that each situation should be considered individually.[‖]

[*] For the Catholic Church's view on abortion, see http://www.newadvent.org/cathen/01046b.htm
[†] For the Methodist view, see http://www.methodist.org.uk/static/factsheets/fs_abortion.htm
[‡] For the Lutheran view, see http://www.lcms.org/pages/internal.asp?NavID=2121
[§] For more information, see http://www.religioustolerance.org/jud_abor.htm
[¶] For more on the Hindu view, see http://www.angelfire.com/mo/baha/hinduism.html
[‖] For more on the Buddhist view, see http://www.urbandharma.org/udharma/abortion.html

In the United States abortion remains legal, the result of the Supreme Court decision in 1973 in a case known as *Roe v. Wade*. However, pressure from religious groups has encouraged some states to pass legislation that limits abortion; the federal government continues to prohibit the use of federal funds to pay for abortions. The restrictions that the U.S. Supreme Court has considered not to be in violation of a woman's right to an abortion include:

- Requiring parental notification and/or consent before a minor may have an abortion
- Requiring a waiting period after an initial visit to a clinic before an abortion can be performed
- Requiring counseling before having an abortion

Given the passions ignited by the debate over abortion, there is little chance that the controversy over this medical technology will subside any time soon.

WHERE WE'VE BEEN

Technology has played an important role in the development of medical and scientific technologies. Medical diagnosis is supported by imaging techniques such as x-rays and MRI. Currently, microsurgery is being performed with the assistance of robots, allowing surgeons to work with much smaller incisions and therefore decreasing the risk of infection and speeding recovery time. The clinical laboratory, where blood tests are performed and bacterial cultures are grown, is also heavily automated. In addition to specialized equipment, hospitals and medical practices make extensive use of general computing for record keeping.

Research in the pure sciences also relies on technology. For example, biological research into small organisms requires the use of microscopes. Light microscopes have been available for hundreds of years, but the invention of the electron microscope in the 20th century eased the analysis of previously unseen structures. Modern meteorology uses technology to measure atmospheric conditions and software models to make weather forecasts. The exploration of the cosmos relies on the technology of telescopes; physics research into subatomic particles and transuranium elements uses massive particle accelerators.

Some of the technologies produced by scientific and medical research are controversial, primarily because they violate the beliefs of a number of religions. Such technologies include cloning, stem cell research, and a group of reproductive technologies (assisted conception, contraception, and abortion).

THINGS TO THINK ABOUT

1. Medical imaging technologies—x-rays, CTs, MRIs—are not without their drawbacks. If you were advising a family member which technology to choose to provide an image of a broken limb, which would you choose? Why? Would your answer change if a physician suspected a lung tumor? Why?

2. Consider the uses of cloning presented in this chapter. Which types do you believe should be allowed? Why? Who do you believe should decide what cloning is allowed? Why?

3. Should there be regulations on stem cell research? Why? If you believe there should be restrictions, what should be permitted? Why?

4. In general, should governments take an active role in regulating scientific research? Should governments take an active role in regulating the practical applications of scientific research? Why?

5. Mammograms for women have been shown to diagnose breast cancer earlier than self-examinations. However, a mammogram is an x-ray, which in itself could cause the cancer it is designed to diagnose. How would you balance the

trade-off between finding existing disease versus causing the disease? Should there be a concerted effort to find an alternative diagnostic technology for breast cancer in women or is the current technology sufficient? Why?

6. How safe do you consider robot-assisted surgery? What other issues besides the safety factor might affect someone's decision to submit to a procedure using robots? Would you undergo robotic surgery? Why?

7. U.S. citizens have come to rely on predictions made by the government-supported National Weather Service. Conduct some Internet research to find statistics about the accuracy of these predictions. Given what you found, how much should we rely on those predictions? Why?

8. Some ideas accepted as scientific fact today—in particular, evolution, relativity, and the organization of the cosmos—are technically theories because they cannot be proven conclusively. The time frames and distances involved are such that experimentation to validate the theories is impractical. What do you believe about these theories that the scientific community accepts as fact? What arguments would you make to someone who believes the opposite of what you believe?

9. China and India are facing significant pressure from overpopulation. In addition to the strategies presented in this chapter, what could these countries do to limit their populations? Is better contraception the answer, or is there another technology that would be helpful? Explain the reasoning behind your answer.

WHERE TO GO FROM HERE

Bellis, Mary. "The Invention of x-ray machines and CAT-scans." *http://inventors.about.com/library/inventors/blxray.htm*

Bellomo, Michael. *The Stem Cell Divide: The Facts, Fiction, and the Fear Driving the Greatest Scientific, Political, and Religious Debate of Our Time.* New York: AMACOM/American Management Association, 2006.

Dick, Richard S., Elaine B. Steen, and Don E Detmer. *The Computer-Based Patient Record: An Essential Technology for Health Care.* Washington, DC: National Academies Press, 1997. (See also *http://www.nap.edu/catalog.php?record_id=5306*)

Engleman, Robert. *More: Population, Nature, and What Women Want.* Washington, DC: Island Press, 2008.

Furcht, Leo, and William Hoffman. *The Stem Cell Dilemma.* New York: Arcade Publishing, 2008.

"Galileo gets credit for refracting telescope." *http://www.reviewtelescopes.com/telescope-types/galileo-gets-credit-for-refracting-telescope-28/*

Goldstein, Joseph, Dale E. Newbury, David C. Joy, and Charles E. Lyman. *Scanning Electron Microscopy and X-ray Microanalysis.* New York: Springer, 2003.

Haney, Johannah. *The Abortion Debate: Understanding the Issues.* Berkeley Heights, NY: Enslow Publishers, 2008.

Kalender, Willi A. *Computed Tomography: Fundamentals, System Technology, Image Quality, Applications.* New York: Wiley-VCH, 2006.

Kaplan, Phoebe, Robert Dussault, Clyde A. Helms, and Mark W. Anderson. *Musculoskeletal MRI.* New York: Saunders, 2001.

Keim, Brandon. "Buddhists and Hindus not worried about scientists playing god." *http://www.nytimes.com/2007/11/20/science/20tier.html?_r=1&oref=slogin*

Kuhn, Thomas. *The Copernican Revolution: Planetary Astronomy in the Development of Western Thought.* Cambridge, MA: Harvard University Press, 1957.

Levine, Aaron D. *Cloning: A Beginner's Guide.* Oxford, Oxfordshire, UK: One World Publications, 2007.

Love, Jamie. "The cloning of Dolly." *http://www.synapses.co.uk/science/clone.html*

Monmonier, Mark. *Air Apparent: How Meteorologists Learned to Map, Predict, and Dramatize Weather.* Chicago, IL: University of Chicago Press, 2000.

Roderick, Daffyd. "Doctor's little helper." *http://www.time.com/time/interactive/health/doctor_np.html*

Rothenberg, Mikel A. *Understanding X-Rays: A Plain English Approach.* Eau Claire, WI: PESI Healthcare, 1998.

"Stem cell basics." The U.S. National Institutes of Health. *http://stemcells.nih.gov/info/basics/*

Wythes, Joseph H. *The Microscopist: Or, A Complete Manual on the Use of the Microscope.* Ann Arbor, MI: University of Michigan Library, Scholarly Publishing Office, 2005.

Entertainment and the Arts 11

WHERE WE'RE GOING

In this chapter we will

- Discuss the influence of technology on the visual arts, including the switch to digital content creation.
- Discover how digital technology has fundamentally changed photography.
- Learn about the effect of the Web on the field of graphic design.
- Look at the changes in live and recorded music brought on by the appearance of synthesizers and other digital instruments.
- Talk about how digital music delivery continues to alter the way in which recorded music is produced and sold.
- Discuss copyright issues surrounding the digital delivery of music and video.
- Learn about the use of computer-controlled equipment in filmmaking.
- Consider how technology has affected animation.
- Look at the challenges of film preservation.
- See how digital video has impacted theaters.
- Examine the changes in the delivery and technology of television arising from the switch to digital formats.

INTRODUCTION

Of all the changes technology has brought to modern society, some of the most visible are the changes to art and entertainment that affect the way in which artistic works are created and distributed. There have been significant economic impacts from these changes, some of which are still evolving. In this chapter we look at how artistic content creation and the distribution of artistic and entertainment products have changed. We also discuss the intellectual property issues surrounding some of the new distribution methods along with the way in which digital media have affected some specific jobs.

ART

One of the first nontechnology fields to feel the impact of computer technology was the still visual arts. Graphics software made it possible for artists to use media other than pencil or paint and paper to produce creative works. Known as *digital content creation*, the generation of artistic works by computer continues to grow. Photography has also been impacted in several ways by technology through the presence of digital cameras. In this section we will look at both digital content creation and photography. We will then consider the relationship between graphic design and the Web.

Digital Content Creation

Digital content creation arose to meet two major needs:

- Publishers who moved to desktop publishing wanted illustrations produced in digital form. Scans of paper-based art didn't yield the resolution or quality publishers needed. This included newspaper, magazine, and book publishers, along with many corporations that produced brochures, catalogs, and annual reports.
- The Web used a great deal of art, all of which needed to be in digital form and subject to compression to speed up downloads.

However, programs to create digital art predate desktop publishing and the Web. The creation of digital art has one important technological requirement: free movement about the screen with a pointing device of some kind (mouse, stylus, or light pen, for example). Digital art therefore had to wait for the graphic user interface. The first widely distributed art program was MacPaint, which shipped with the first Macintosh in 1984. It used the mouse as a drawing tool to create black-and-white, single-layer paintings.

From that beginning, digital art software has expanded to include color as soon as computers had enough main memory to handle it.[1] Bit-mapped paint-

---

[1] In a black-and-white painting, each dot in the painting requires only one bit of storage. However, when color is added, the colors in the palette must be numbered and then assigned to each dot. To represent four colors, two bits per image dot are needed; for 16 colors, four bits are needed. A minimally acceptable color image uses 256 colors, requiring eight bits per dot, and true-color (thousands of colors) requires 24 bits per dot. Most of today's computers also support 32-bit color (millions of colors). To find out how much storage an image requires before compression, multiply the horizontal and vertical resolution of the image to get the number of dots. Then multiply the number of dots by the bits per dot and divide by 8 to translate into bytes. It is amazing how much space a true color image can take up, regardless of whether it's in main memory or on disk.

ing images were joined by object graphics, in which the elements of the drawing were represented by shape, color, size, location, and position in the layering of elements in the image. Object graphics, which could be rendered by laser printers, generated better output than the paintings, which had to be drawn using individual dots. The first object graphics programs included MacDraw, Macromedia Freehand, and Adobe Illustrator. Only Illustrator survives today, however.

Graphics programs designed for artist works have matured since 1984, along with output devices that can print high-resolution bit-mapped images that look as good as those produced by object graphics. Color printers, although still evolving, can provide decent output at reasonable prices.[2]

One of the problems artists cited with using computers to create artistic works was that a mouse wasn't a natural tool for an artist. The first technology that attempted to address that issue was the graphics tablet, on which an artist draws with a stylus (**Figure 11.1**). Even a graphics tablet, however, doesn't simulate the way in which an artist would work because the drawing appears on a screen rather than under the stylus. Technology has answered this problem with displays, such as that in **Figure 11.2**, on which an artist can draw. The modifications the artist makes to a work of art appears directly under the artist's stylus, as if he or she was working with canvas or paper.

Digital content creation has not put artists out of work, nor has it affected the market for fine arts. Digital art tends to have a different purpose from hand-created artwork. Although we may hang prints (reproductions of fine art) or computer-generated images on our walls, the demand for original works of art in traditional media appears to be both consistent and solid.

Photography

Film photography began in 1826, with the first permanent photograph taken by Frenchman Nicéphore Niépce. His image required an 8-hour exposure. From that time through to the late 1980s, people created photographs with silver nitrate–

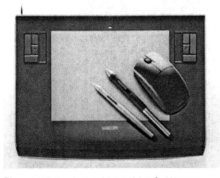

Figure 11.1 A graphics tablet (a Wacom Intuos 3) with stylus and cordless mouse.

Figure 11.2 The Wacom Cintiq display, on which an artist draws with a stylus.

[2]There's an old saying in the printing and publishing industry that "color costs." When color is added to any type of output, it costs more than black-and-white or gray scale. At least for the moment, that continues to be true.

based film. The appearance of digital cameras in 1989, however, began a change that has nearly eliminated film photography.

The first well-known digital camera was the Sony MAVICA.[3] It used a 2-inch floppy disk to store images, which were captured by two charge-coupled devices.[4] Each image was represented by 720,000 pixels. The floppy disks stored either 25 high-quality images or 50 lower quality images. The quality, however, was far below that of the film cameras then in use.

Since the introduction of the MAVICA, the number of pixels per digital image has continued to rise. At the high end, the Hasselblad H3D11-39 (**Figure 11.3**) generates images that are 7,212 pixels by 5,142 pixels (approximately 39 megapixels). Like most high-end models, it is a professional single-lens reflex with quality equal to or better than professional film cameras. At the time this book was written, however, such cameras were extremely expensive: The Hasselblad H3D11-39 sells for about $34,000 or can be rented for about $500 a day. Digital cameras for the rest of us produce images of up to 10 megapixels and generally sell for under $500.

Instant photographs have long been popular, primarily in the form of Polaroid's self-developing film. It's not surprising, then, that consumers were quick to adopt digital cameras, especially in conjunction with low-cost printers that produced output similar to that of lab-printed photographs and cameras appearing in all sorts of devices (cell phones, personal digital assistants (PDAs), and computers of all types and sizes). Consumers were so heavily in favor of digital cameras that in February 2008 Polaroid announced that it would be shutting down its instant film camera division and licensing the technology to some other firm that wanted to continue to provide the film for existing instant film camera owners.[5] Another indication of the switch to digital photography is that by 2003 more digital cameras were sold than film cameras[6] and by 2006 Kodak was selling more digital products than film.[7]

Figure 11.3
The Hasselblad H3DII-39 camera, which produces images of up to 39 megapixels.

---

[3] MAVICA is an acronym for MAgnetic VIdeo CAmera.

[4] Charge-coupled devices were developed at AT&T Bell Labs in 1969. Their purpose is to translate light into electrical signals that can be stored as digital data.

[5] For details, see http://www.boston.com/business/technology/articles/2008/02/08/polaroid_shutting_2_mass_facilities_laying_off_150/

[6] See http://query.nytimes.com/gst/fullpage.html?res=9B03E3DE163EF936A15751C1A9659C8B63

[7] See http://www.dpreview.com/news/0601/06013102kodaksales.asp

In contrast, professional photographers continued to use film-based cameras until the quality was nearly equal to that of film cameras. Today, some professional portrait studios are totally digital. They use software such as Photoshop to retouch portraits rather than doing it by hand; they can provide instantaneous proofs to customers on a computer monitor. Because the quality of the images rivals the film cameras previously used in portrait photography, customers are satisfied and the studio's costs go down (after the expensive camera has been paid for!). Professional photographers have not abandoned film entirely, however. Some believe that at low film speeds, even 20-megapixel digital photography isn't as good as film and that at high film speeds film is better than cameras with low resolution.[8] A general consensus seems to be emerging, however, that digital cameras become better than 35-mm film at about 11 megapixels.[9]

Just as some audiophiles believe that vinyl records provide better fidelity than compact disks (CDs), it seems likely that there will always be some photographers who prefer film and that there will always be film sold to service those customers.[10] Nonetheless, it does appear that the bulk of photography is digital rather than on film.

The switch to digital photography has had more than just an economic impact on the film and camera markets. It has become relatively easy to alter a digital photograph and extremely difficult to detect a good alteration once the file has been saved. Photographs can therefore show anything a photographer or artist wants them to. We can no longer trust that a photograph presents a true picture of something. For most people this is not an issue, but for fields such as forensics, it becomes of vital concern. Assume, for example, that a driver hits a pedestrian in a crosswalk. To protect himself, the driver takes a digital photo of the accident scene. Once home, he alters the photo by moving the lines of the crosswalk. The victim was actually in the crosswalk when the accident occurred, but the photo that the driver shows to the police shows that the crosswalk was further down the street and that the victim was not in the crosswalk at all. Who should the police believe, the victim or the photograph?

Graphic Design and the Web

Just about anyone can create a Web site. However, it takes more than being able to use Web development software to create effective and well-designed Web pages: It takes knowledge of graphic design.

The beginning of the Web made it clear how easy it is to create something with a poor design. The first people to create Web pages had little or no training in graphic design, and it showed. The pages had the default gray background, no text styling other than the default HTML styles, and sections separated by horizontal rules. Many of these pages were created before tables became available to

---

[8] For research into the relationship between film speed and digital resolution as compared with film, see http://www.clarkvision.com/imagedetail/film.vs.digital.summary1.html

[9] See http://www.normankoren.com/Tutorials/MTF7A.html

[10] Along the same line, there are audiophiles who believe that amplifiers with tubes rather than integrated circuits provide a richer sound. Replacement tubes are available—although expensive—for those who own this type of high-end equipment.

give developers at least a minimum of control over the spacing of design elements. The situation has improved to some extent because of two factors. First, HTML has improved to the point where it can give Web page creators considerable control over what appears on the page. Second, Web design software makes it possible for developers to create sophisticated page layouts without writing HTML code.[11]

Good design does matter, for several reasons (assuming the goal is to get as many people as possible to visit a site and to return to it many times):

- Poor design offends the artistic sensibilities of some users and keeps them away from the site.
- Poor design for the navigation of a site makes it difficult for users to find the information they need. If a site is trying to sell a product and searching the catalog is not intuitive, for example, users won't find the products they want and therefore will purchase less than they might otherwise.
- Some design elements—in particular, flashing text and/or graphics—can have negative physical consequences, especially on those with epilepsy.
- Text that is too small can make it impossible for users to read the site. They won't be able to get the information the site operator wants them to have, whether the site is selling a product or providing information.
- A poorly designed Web site can give a potential customer a negative impression of a company and prevent a potential customer from becoming a real customer.

Given that the design of a Web site can significantly influence its effectiveness and the reputation of the organization it represents, graphic designers are in as much demand today as they were before the appearance of the Web. What it does mean, however, is that the job of the graphic designer has changed (or, perhaps better, expanded) to meet the needs of the new venue. Graphic designers, who traditionally worked with paper layout and typography, now find themselves working with designing for online viewing. Like so many other fields that we have discussed in this book, graphic design has been changed by technology. In this case, it's not necessarily for the worse, given that employing a graphic designer can significantly improve an organization's Web site.

To meet the need for online graphic designers, college majors in digital media, which are typically housed in art departments, have been established to prepare graphic designers to work with computer-based tools. Course work also includes classes in traditional graphic design with emphasis on the differences between print and online media. This field is distinct from the Web development curriculum that is part of a college's Information Technology major, which focuses on the programming side of Web pages.

---

[11] Poor Web designs have not disappeared. Consider, for example, the site at http://www.reading.ac.uk/GraecoAegyptica/. The background isn't gray, but the underlying graphic makes the first-generation text nearly impossible to read. For links to more poor designs, see http://www.webpagesthatsuck.com/dailysucker/

MUSIC

The music field has experienced a significant impact from the introduction of technology. Not only do we now download much of our recorded music over the Internet, but digital instruments play a significant role in music creation. In this section we look at the technologies behind digital instruments and how they have affected musicians. Then we turn to the still-emerging area of digital music distribution.

Synthesized Instruments

We often think that computerized instruments were developed in the 1950s, beginning with the Moog synthesizer. However, people have been using technology to power instruments since the second century B.C., when a Greek invented a hydraulic mechanism to power an organ. The aeolian harp, which made sound when wind moved across its strings, was developed during the same period. **Table 11.1** summarizes early machines that assisted in making music.

Today's synthesizers can either create sounds that mimic acoustic instruments or create sounds that only can be generated electronically. In addition, electronic instruments can "sample" sounds, a process through which the instrument takes a digital reading of a sound at regular instruments. When the samples are played one after the other, the synthesizer outputs a sound close to the original.[12] Some of the finest electronic instruments, such as the Yamaha Clavinova, which feels and sounds like a grand piano, produce all of their sound from sampled analog sources. The Clavinova has a full-sized 88-key keyboard along with electronics for sampling, storing, and playing digital sounds.

In addition to totally electronic instruments, many musicians today use acoustic instruments that have been fitted with electronic pickups that send the sound to amplifiers and ultimately to speakers. The reaction of musicians and music aficionados depends to some extent on whether the instrument in question is completely electronic or whether it is amplified.

The first mechanical instrument to draw the scorn of musicians was the player piano. A player piano is an acoustic piano in which the keys are moved mechanically and the sound is therefore true to human piano playing. Modern player pianos work in the same manner, even though they may include electronics to play other digitized sound. The primary argument leveled against player pianos was that although they could play the notes of a piece of music perfectly, they were unable to reproduce the "soul" that a pianist puts into a performance. Player pianos are often fun to have and can fill in where a pianist is not available. However, even today people flock to concerts to hear musicians play the piano (regardless of the type of music). The player piano seems to have had minimal effect on keyboard players.

---

[12]How well sampled sound matches the original depends largely on the "sampling interval." The closer together the samples, the closer the digitized sound matches the original.

Table 11.1 Early Mechanical Instrument Development

Date	Instrument
2nd century B.C.	Hydraulis, a hydraulic assist for an organ
	Greek aeolian harp, which made sound when wind passed over its strings
15th century	Development of the hurdy-gurdy
16th century	Mechanical organs
	Water-assisted organs
	Archicembalo invented, using six keyboards and a 31-step octave
17th century	"Nouvelle Invention de lever," a hydraulic device to make sound
18th century	Mechanical singing birds commonplace
	Barrel organs in use
1761	Clavecin electrique, developed by Abbe Delaborder in Paris
	Glass harmonica, developed by Benjamin Franklin
	Panharmonicon, developed by Maelzel
1796	Development of the carillon
	Development of the music box
1867	Electromechanical piano, developed by Hipps in Switzerland
1876	Electroharmonic piano, developed by Elisha Grey; sent musical tones over telephone wires
	Tonametric, developed by Koenig; provided four octaves divided into 670 parts
1887	Development of the player piano in the United States
1895	Octavia (producing 8th tones) and the arpa citera (producing 16th tones), developed by Julian Carillo
1880s	Dynamaphone, or telharmonium, invented by Thaddeus Cahill The 7- to 200-ton instrument used alternating current through dynamos The telharmonium is considered to be the first real music synthesizer
	Choralcello, developed by Melvin Severy and George B. Sinclair; an organ that generated sound with an electromagnetic tone wheel
1899	Singing arc, the first all-electronic instrument, debuts. The device, invented by William Duddell in England, used the sounds emitted by a carbon arc lamp. The performer varied the sounds using a keyboard.

Date	Instrument
1920	Aetherophone, developed by Leon Theremin in Russia; used vacuum tubes for synthesizing a beat. A performer changed the pitch by moving a hand up and down a rod. The device became known as the Theremin.
1926	Spharophon, the Partiturophon, and the Kaleidophon, developed by Jorg Mager in Germany; electronic instruments in theaters
1928	Ondes Martenot (developed in France by Maurice Martenot) and the trautonium (developed in Germany by Friedrich Trautwein) improve on Mager's instruments. Performers use devices such as a ring passing over a keyboard to change the pitch.
1929	Hammond Organ Company is founded by Laurens Hammond.
1930	First drum machine appears.
1932	The first electric guitar is used in a concert.
1937	Development of the first electronic-acoustic piano
1948	The first instrument able to synthesize the sound of a real instrument—a cello—is perfected by Hugh LeCaine, a Canadian.
1940s	Development of a number of small electric organs
1949	Robert Moog builds his own Theremin.
1950s	Concerts using the Theremin are still occurring.
	People use various means to generate sounds electronically.
	First protests by musicians that those producing electronic music were performing but hadn't joined the musicians' union.
1955	Electronic music synthesizer, developed By Olson and Belar. This device was programmable from a keyboard.
1958	At Bell Labs, computers are first used to generate music.
	Bell Labs researchers write computer programs to generate music.
1962	The synket is built by Paul Ketoff as a performance instrument.
1964	The Moog synthesizer debuts.
1970s	Concerts using the Theremin are still occurring.

The story with all-electronic instruments is somewhat different. The development of electronic keyboard instruments that could synthesize the sound of multiple orchestral instruments, along with the sophistication of the drum machine, brought fears that studio musicians and show orchestras would become a thing of the past. Such electronic instruments have been used both for recorded music and to accompany live stage shows, but regardless of how good the synthesizers may be, many listeners can still discern a difference between the machines and live performers. The feared drop in musician employment has not occurred. Instead, electronic instruments have found their place in rock, modern, and experimental music, along with human musicians.[13]

Digital Music Delivery

At the time this book was written, the changes in the delivery of recorded music—specifically, the emergence of online delivery—were garnering a great deal of attention. Music is an ideal item to be transferred electronically: The files are relatively small and the sound fidelity is preserved whenever the file is copied. However, most recorded music is also under copyright and, as you are likely aware, the issue of illegal copying of copyrighted material continues to loom over the music industry.

Changes in the Music Delivery Process To understand the impact of digital music delivery, we first need to consider the traditional way in which recorded music has been sold. The process goes something like this:

1. An artist is signed by a record company.
2. The artist goes into the recording studio and produces the music that will be sold. The artist usually must record a collection of material, enough to fill a CD.
3. The record company arranges for production of the media that will be sold, including contracting for cover art.
4. The physical media are produced and shipped to distributors.
5. The distributors sell the media to record stores and then ship the product to the stores.
6. The store sells the media to the consumer.

Notice how many organizations are involved in getting the recording to the consumer (five: artist, recording company, media duplication company, distributor, and retail outlet). Each needs to be paid in some way. The recording company is at the top of the hierarchy. It collects money from distributors and uses it to pay for media duplication and artist royalties. Whatever is left after promotion expenses is profit for the record company.[14]

---

[13]For a level-headed analysis of the impact of electronic instruments, see "Should One Applaud?" by Trevor J. Pinch and Karin Bijsterveld (*Technology and Culture* 44 (2003): 536–559).

[14]Independent artists can eliminate the record company and produce their own recordings for sale. However, they traditionally have not had the capital to promote their recordings as widely or as vigorously as can recording companies that have more money and large marketing departments.

Digital music delivery changes the process, often eliminating the recording company and almost always eliminating the production of physical media. Because no CDs are being produced, the artist or recording company eliminates that cost. However, the cost of producing a CD, even with full-color cover art and a jewel case, is less than $1 when produced in large quantities. The savings to the consumer from the lack of physical media alone are negligible.[15]

However, digital delivery of music can also eliminate the distributor; retailers (for example, iTunes) or recording companies can sell directly to the consumer. This means one less entity involved in handling the recording, which eliminates one "mark up" of the price of the product. Because the distributor's costs are substantial—shipping, storage facility expenses, marketing, and so on—eliminating the middleman can noticeably reduce the cost to consumers.

The change to digital music delivery has had a significant impact on CD sales. In the first three months of 2005, for example, CD sales dropped 20%. Consumers also tend to purchase individual tracks rather than entire albums.[16]

Impact on the Artists Musicians, for the most part, seem to like the digital delivery of music. It opens up the entire Internet as a market for independent artists and, because of lowered costs, offers the *potential* of higher royalty rates to those who do work through a company.[17] In addition, digital music delivery has provided some previously unavailable opportunities:

- Artists can allow users to sample an album that will be appearing on CD by giving away a track or two online.
- New artists can distribute their music to a wide audience via the Internet and potentially build demand for their work.
- Artists can give away entire recordings to build demand for concerts.
- Artists no longer need to record an entire CD. Instead of filling a CD with work an artist may not consider to be the best, the artist can record only a few tracks and sell them individually. This allows an artist to experiment at low cost (just the cost of the studio time).
- Artists can make recordings independent of a record company and sell them without using even a distributor, meaning that they keep more of the money that sales of their product generates.[18]

At the same time, there are risks for an artist when turning to digital music distribution. In particular, the artist is vulnerable to the honesty of the vendor selling music online. There is no simple way to audit digital sales to determine whether all royalties due have been paid.[19]

---

[15] The economics are a bit different for an independent artist, who must foot the bill for the production of CDs. A small order (500–1,000 units) can run about $6.50 per CD. (See http://www.eqmag.com/article/the-future-music/Sep-06/22874 for details.) Digital distribution can therefore save the independent artist a lot.

[16] See http://www.breitbart.com/article.php?id=070322121539.enwwmbqh&show_article=1

[17] Note the word "potential" in this sentence. Although higher royalty rates in the presence of lower costs make sense, it hasn't happened.

[18] Although recording company contracts pay royalties, in many cases the artists are responsible for covering expenses such as studio costs. The recording company advances the money and then repays itself out of the artist's royalties.

[19] For a jazz musician's take on this issue, see http://www.allaboutjazz.com/php/article.php?id=21697

Copyright Issues As soon as people were able to create digital recorded music, music was shared freely over the Internet. Services such as Napster arose to make it easy for Internet users to exchange files. Napster was a peer-to-peer file-sharing service. It did not actually store music files but instead let users know where specific recordings could be found. Users then downloaded directly from other users.[20]

As far as the recording industry was concerned, peer-to-peer sharing of music was copyright infringement and outright theft. Their reaction was to take legal action against Napster. However, Napster claimed that it was not distributing music files and therefore could not be held liable for how its consumers used the service. Despite its claims to the contrary, Napster was found guilty of illegal music distribution and was ostensibly forced to change the nature of its business. (The current version of Napster still supports peer-to-peer access to music, but the service [which now costs money] allows users to listen to music but not download it.)

Napster's absence in peer-to-peer music sharing was filled by a variety of other similar services, most notably Kazaa. By 2006 Kazaa had been demonized as "...an international engine of copyright theft which damaged the whole music sector and hampered our industry's efforts to grow a legitimate digital business."[21] Within a year Kazaa, like Napster, had converted to a legal form of business.

Digital delivery of music has evolved into a paid, legal service. The first, and largest, digital music seller is Apple's iTunes, which also handles video (movies and television).[22] Apple negotiated agreements with recording companies for the legal sale of digital recordings. However, the copyright issue didn't go away. All music tracks were sold initially with Digital Rights Management (DRM) software that, in this case, prevents the use of iTunes products on any MP3 player other than an Apple iPod.[23] Given that the iPod held the largest share of the MP3 player market and that the music was inexpensive (generally less than $1 a track and $10 an album), the restrictions placed by the presence of DRM didn't appear to be a hindrance.

By 2007 some users were chafing under iTunes's DRM restrictions. Apple therefore negotiated with some recording companies to offer DRM-free music. Initially, the DRM-free tracks cost a bit more than those with the copy protection ($1.29 vs. $.99), but by the end of 2007 the price of the copy-protected and non–copy-protected music became the same at the original price.

The existence of legal sources for digital music does not mean that copyright infringement issues have disappeared. Two organizations—IFPI[24] (international, covering 1,400 recording companies in 70 countries) and RIAA[25] (U.S.)—work constantly to protect what they consider to be the rights of artists and recording companies.

---

[20] For details on the Napster issue, see http://iml.jou.ufl.edu/projects/Spring01/Burkhalter/Napster%20history.html
[21] John Kennedy, CEO of IFPI, quoted in http://www.mipi.com.au/statistics.htm
[22] As of February 2008, iTunes was the second largest music retailer in the United States, following Wal-Mart.
[23] The iTunes license allows the copying of the music to be played on multiple computers and the creation of CDs from the music files. The restriction is on MP3 players only.
[24] http://www.ifpi.org
[25] The Recording Industry Association of America, http://www.riaa.org

The RIAA is well known for its legal actions against those it suspects of illegal downloading. In February 2008 it sent letters to individuals at 12 major universities, threatening legal action for illegal downloading. Those who received the letters had to either settle outside of court, paying a fee to cover the illegal downloaded music, or become the target of legal action.[26] RIAA's reputation has suffered to some extent from its legal activities because some of the targets of its suits and the amount of the settlements it requested have been considered outrageous. Parents have been threatened with lawsuits as the result of downloading by their children; fines of more than $10,000 per recorded track have been requested from individuals, some of whom were children.[27] The RIAA has also come out against the legality of copying the contents of legally purchased CDs to an MP3 player.[28]

Exactly what constitutes legal uses of recorded material in the United States is governed by the Digital Millennium Copyright Act (DMCA) of 1998.[29] The law attempts to define what is considered "fair use" of copyrighted material.[30] Its major provisions include the following:

- It is illegal to disable any copy protection placed on digital content.[31]
- It is illegal to create or sell software or hardware that is designed to be used for breaking copy protection.
- Exempts IPSs from liability if their services are used for illegal transfer of copyrighted material but requires them to remove illegal copyrighted material from their servers when notified of its presence.
- Exempts colleges and universities from liability if their networks are used by faculty, staff, or students to copy copyrighted material.
- Establishes that those who broadcast copyrighted material over the Internet must pay royalties, much in the same way that performers pay fees to the American Society of Computers, Authors, and Publishers (ASCAP) for live performances of copyrighted works.

The DMRA was backed by the software, movie, and recording industries but has been opposed by libraries, universities, and other similar bodies. In particular, researchers in many subject areas believe that it has blocked their "fair use" of copyrighted material in research.[32] Some of the impact was mitigated by a provision of the law that stated that the U.S. Copyright Office was to watch for negative impacts of the law on academic and research activities and to provide exemptions in such cases. Exemptions due to expire in 2009, for example, allow the circumvention of copy protection in cases where outdated technology prevents

[26] For details, see http://www.riaa.org/newsitem.php?id=B0FAEEC1-A56A-0F04-D999-94A807ADAA6E. For a university's reaction, see http://www.ur.umich.edu/0304/Jan26_04/04.shtml

[27] For one side of this argument, see http://www.newscientist.com/article/dn3877-internet-musicsharers-face-legal-attack.html

[28] http://www.eff.org/deeplinks/2006/02/riaa-says-ripping-cds-your-ipod-not-fair-use

[29] The text of the DMCA can be found at www.copyright.gov/legislation/dmca.pdf

[30] The DMCA applies to software as well as digital music.

[31] The exceptions to this provision include activities such as the conduction of encryption research and testing computer security.

[32] Previous copyright laws have considered the use of small portions of copyrighted material for research, teaching, and reviewing as "fair use."

the use of software or computer games. However, the Copyright Office turned down a request for an exemption for format shifting, where someone takes a legally owned copyrighted item and moves it to another medium.[33]

As the delivery of recorded music continues to shift from physical media to digital downloading, the issues of copyright will almost certainly continue to be problematic. The transition to digital distribution is not complete; it would appear that there has not yet been time to achieve an acceptable balance between the rights of those who produce the music and the rights of those who buy it.

FILM

Like music, film has gone digital. Videos are now sold online through the same stores that sell recorded music; they are also tangled in the same copyright issues. We've discussed the copyright issues in the context of music and therefore won't cover copyright again. However, technology has also had a major impact on the way in which films are made, in particular in the areas of animation and special effects. We also look at the need for film preservation, which has only become as issue recently as films from the early part of the 20th century begin to deteriorate.

Computer-Controlled Filmmaking

Computers entered filmmaking not through animation but through the technology of cameras. In the beginning, many special effects were created using models and stop-motion photography. Whenever a film needed a "creature," a model maker sculpted and painted a model that could be moved. Filming using the model required moving the model, filming it, then moving it a small bit for the next frame, and repeating the process. Early science fiction films, such as Ray Harryhausen's work for *The 7th Voyage of Sinbad* (1958), used that technique heavily.

When a film involving stop-motion photography required more than one element—for example, human actors and creatures—the images were prepared separately and then filmed again to bring them together. The more images to be combined, the more difficult it became to align them accurately. This is where computers came in.

One of the first films to use computers to record camera actions was *Star Wars: Episode IV*. Engineers at Lucasfilms developed cameras that could "remember" the sequence of positions that were used to record a specific scene. For example, the filming of the fighter run through the trench of the Death Star at the end of the film used a model of the trench and models of the fighters. Shooting the scene required multiple passes of the different fighter models down the trench. To obtain film that could be combined correctly, computer-controlled arms moved the models down the trench. The movements of the models and the cameras that filmed them could then be given as inputs to the cameras and the arms moving the models to create another version of the scene where the models moved in the same path and the cameras used the same angles as the previous shot.

---

[33]For more information, see http://www.sourcewatch.org/index.php?title=Digital_copyright

The Changing Technology of Animation

Animation began as long ago as 1892. The first movies were animated shorts using loops of pictures. However, it wasn't until 1906 that an animated movie was made on film. By today's standards, the first efforts were crude (black-and-white line drawings). The first true cartoon appeared in 1914 featuring a likable dinosaur named Gertie.[34] By the time Disney released the cartoon *Steamboat Willie*[35] in 1928, animation was an accepted art form.

Computer-Generated Animation   Once upon a time animated films were created by artists who painted each frame on plastic (a "cel"), which was then painted by hand. It took literally thousands of individual, hand-drawn illustrations to create a single film. The cels were photographed and then displayed rapidly to create the illusion of motion, just like the images in a child's flip-book.[36] Because each cel was drawn individually, the process was time consuming and very labor intensive. Many animated films and cartoons were created using cel animation, including the Disney films such as *Snow White* and *Peter Pan* that we consider to be classics.

The introduction of computer graphics changed animation profoundly. The Disney film *Tron*[37] (1982) is generally regarded as the first film to rely heavily on computer graphics. Although there are some live action sequences in the film, most of the story takes place inside a computer. To today's viewers, the film looks as if the scenes inside the computer are computer generated. However, they are really live action shots enhanced by colorization and "rotoscopic" techniques[38] to make them appear like what audiences would think was computer-generated graphics.

One of the first films to use computer-generated animation rather than models for special effects was *The Last Starfighter*[39] in 1984.[40] The story involved a boy whose video game skills brought him to the attention of an alien star force that recruited video game players as pilots. Images of the starfighter spacecraft and space battles were totally computer generated and spliced in among the live action sequences. It was very clear to viewers which scenes were shot on film and which were computer generated; no attempt was made to integrate the animated action with live action.

Today, computer-generated animation can be used to create three-dimensional characters that are so real we sometimes forget they are animated. The first animated film featuring the realistic animation was *Toy Story*[41] (Disney, 1995). It was followed by films such as *Shrek*[42] (DreamWorks Animation, 2001) and

---

[34]To see the entire *Gertie the Dinosaur* cartoon, go to http://www.youtube.com/watch?v=UY40DHs9vc4

[35]Watch *Steamboat Willie* at http://www.youtube.com/watch?v=uKm41_LMPRM

[36]It requires 24 frames per second to create the illusion of movement. A 90-minute animated film therefore requires 21,360 frames. In addition, some frames consisted of multiple cels laid on top of one another, for example, to combine characters and backgrounds.

[37]You can find clips from *Tron* at http://www.youtube.com/watch?v=eMY1FaOKF-0 and http://www.youtube.com/watch?v=41hMgB-ptkw&feature=related

[38]Rotoscoping, a technique in which animators draw on top of film of live actors, has been in use since 1915.

[39]You can view clips of *The Last Starfighter* at http://www.youtube.com/watch?v=poUQIFCQwlk&feature=related and http://www.youtube.com/watch?v=JFOHy791kRE&feature=related

[40]See http://www.imdb.com/title/tt0087597/ for more information.

[41]The theatrical trailer can be found at http://www.youtube.com/watch?v=DPMvfaF2tao

[42]You can find the first theatrical trailer for *Shrek* at http://www.youtube.com/watch?v=jxqQPrUomTc

Cars[43] (Disney, 2006). So many high-quality animated films were being released that in 2001 the Academy of Motion Picture Arts and Sciences created a new award category for those films.[44] We have now come to expect animated films that are visually realistic and that employ A-list voice talent.

Combining Computers and Live Action The combining of live action and animation has a long history. As early as 1949 studios were able to overlay animation created using the traditional cel technique onto live action film. For example, the last minute of the MGM cartoon *Senor Droopy*[45] shows an actress with the cartoon character Droopy on her lap. She appears to be petting the dog's head. Gene Kelly danced with Jerry the mouse in *Achors Aweigh*,[46] and the Disney film *Mary Poppins*[47] used the same method to show stars Julie Andrews and Dick Van Dyke dancing with penguins and other fanciful animals. These early efforts, however, had minimal interaction between the live action and animated characters.

Many of us were awed, therefore, when in 1988 a film called *Who Framed Roger Rabbit*[48] appeared using a combination of animation and live actors integrated so smoothly that it appeared the animated characters were in the real world. This film was only the beginning of what has grown into the sophisticated combination of live action and animation.

In some cases human actors perform roles during filming wearing suits covered with sensors that record movement. Later, the movement is used as a template for the way in which an animated character will move. When the animation is complete, it is inserted into the film in place of the live actor. Golum, from *Lord of the Rings*, and Jar Jar Binks, from the first *Star Wars* trilogy, were created in this way. The technique is known as "motion capture."[49]

Motion capture has another major advantage over using totally animated characters. When purely animated characters are used, the live actors must perform their roles without the benefit of the character that is to be added later. They must nonetheless place themselves in space as if the animated character were present, something that is difficult to do and therefore may create an awkward result.

One of the concerns with the use of computer-generated characters in place of live actors has been that roles for live actors—which often cost a great deal more than animation—will decrease. Although there has been an increase in the number of animated films produced, there has not been a concomitant reduction in roles for humans. All the animated characters require voice actors. Often, the voice actors are filmed as they record the dialog. The animators can then use the facial expressions of the voice actors to make the animated characters appear more real.

---

[43] View a *Cars* trailer at http://www.youtube.com/watch?v=JzwWqkxBb5I&feature=related
[44] The Oscar winners for best animated feature film include *Shrek* (2001), *Spirited Away* (2002), *Finding Nemo* (2003), *The Incredibles* (2004), *Wallace & Gromit* (2005), *Happy Feet* (2006), and *Ratatouille* (2007).
[45] You can watch the entire Senor Droopy cartoon at http://www.youtube.com/watch?v=RJahbFhKfJE
[46] See the dancing scene at http://www.youtube.com/watch?v=BTXgejy5WuI
[47] You can find the dancing with penguins scene at http://www.youtube.com/watch?v=cloJX9KXjaA
[48] The first five minutes of the film can be found at http://www.youtube.com/watch?v=WswD1djENJE. Watch through until the end to see the integration of live action and animation. The trailer can be viewed at http://www.youtube.com/watch?v=zG37ysSqgq8
[49] For a great example of the relationship between motion capture actors and the resulting animation, see http://www.youtube.com/watch?v=EDbMwxs6w9A

Film Preservation

Movies have been around for barely 100 years, but the film on which many early films were shot is deteriorating. Movies—feature films, documentaries, and even newsreels—are in danger of disappearing. Films of all types are a mirror of the times in which they were made, reflecting the attitudes of their makers toward the world around them. Films are therefore worth preserving not only for their artistic merit but also as historical documents. Martin Scorsese, the feature film director, stated: "Film is history. With every foot of film that is lost, we lose a link to our culture, to the world around us, to each other, and to ourselves."[50]

According to the Library of Congress, less than 20% of the U.S.-made films from the 1920s survive; for films before 1910 less than 10% survive.[51] The problem is the nature of film itself. Films face destruction from nitrate deterioration and color fading. The "vinegar syndrome," caused by the deterioration of the film (rather than its content), leaves film with a characteristic vinegar smell and acetate that crumbles and ultimately disintegrates.[52]

The National Film Preservation Foundation suggests three strategies for preserving film:

- Printing old film onto new, more stable film stock.
- Storing film materials under cool-and-dry conditions.
- Providing access through modern copies.[53]

Regardless of which of the three strategies for film preservation an organization undertakes, the preservation is costly. Much of it has been undertaken by the major film studios, which are acting to preserve their own film archives. However, there are many films that do not come from major studios. These "orphan" films are being preserved by other organizations with ties to the film industry, such as the previously mentioned National Film Preservation Foundation, the National Film Preservation Board (an arm of the U.S. Library of Congress),[54] the American Film Institute,[55] and Film Forever.[56]

Digital Video

Traditionally, feature films have been shot on film, developed, duplicated, and sent to theaters. Although many television shows have been recorded on tape, feature films never used that medium. The development of digital video cameras, however, opened up a new way to record video.

Amateur videographers have adopted digital video. It is both easy and convenient to record home video on the memory card in a digital camera and then transfer it to a PC for editing, viewing, and burning to a DVD. To many consumers, digital video is a logical extension of digital still cameras.[57]

---

[50] http://www.pbs.org/newshour/bb/media/jan-june01/orphanfilms_01-12.html
[51] http://www.loc.gov/film/plan.html
[52] For detailed information on these and other dangers facing film, see the Preservation Basics section at http://www.filmpreservation.org/
[53] http://www.filmpreservation.org/preservation/why_preserve.html#
[54] http://www.loc.gov/film/
[55] http://www.afi.com/about/preservation/aboutpresv/deteriorate.aspx
[56] http://www.filmforever.org/
[57] A number of digital video cameras can also take still photographs, although the resolution typically isn't as high as that available with similarly priced still digital cameras.

Given what has happened to still photography, it would seem logical that digital video for feature films would replace the use of film. Digital films for theaters bring several advantages:

- Digital duplication is cheaper than producing film.
- It is cheaper to distribute digital movies. They can be broadcast to theaters via satellite (eliminating duplication costs entirely) or shipped on small, lightweight, inexpensive optical media.
- Digital movies don't decay or need to be preserved as do film movies.

On the other hand, theaters have significant investments in film projection equipment, and the change to digital equipment is often prohibitively expensive. The switch to digital projection has therefore been slower than originally expected. For example, in 2005 less than 200 theaters in the United States had digital projection capabilities; by the end of 2007 that number had risen to around 2,500. The projected growth for 2008 suggested a doubling of the 2007 total.[58] This rapid growth doesn't seem to be as impressive, however, when we consider that there are just over 12,000 theaters in the country.[59]

TELEVISION

Television shows are sold today in the same manner as feature films. You can purchase them on tape, on DVD, or in digital format that is downloaded over the Internet. You can also view television shows from Web sites such as joost.com. However, that isn't the only major change that has come to television in the past decade: The switch to digital audio and video is still in progress.

Receiving Programming

Initially, television broadcasts came over the air. If you didn't live in an area that could receive over-the-air transmissions, you were simply out of luck. In 1948, however, Robert Tarlton created the first cable television system by installing a large antenna on a hilltop and then sending the signal to his Pennsylvania community over a coaxial cable. Cable television was followed by analog satellite television, beginning in the late 1960s, with significant programming available in the mid-1970s. The large analog C-band dishes largely became outdated with the appearance of digital satellite television in the early 1980s.[60] The small, "direct broadcast" satellite systems that we commonly use today are much cheaper than analog satellite setups and are competitive in price and programming with cable television.

---

[58] See http://www.signonsandiego.com/uniontrib/20080113/news_1a13digital.html for more information.
[59] The source of the total number of theaters is http://www.statemaster.com/graph/lif_cla_mov_the_and _dri-classic-movie-theaters-drive-ins
[60] One difference between the large dishes and today's small dishes was the way in which programming was sold. Users of small dish service have to purchase a complete package of channels, just as in cable television. However, C-band channels could be purchased a la carte; a user could create his or her own programming package. In addition, many companies provided the same C-band channels and used the hardware the user had in place. There was direct competition between providers, which tended to keep prices in check. Today users can choose cable or one of only a few satellite providers, each of which has its own equipment.

In the early 2000s, cable television providers began to transmit digital signals along with analog signals. Subscribers with digital television sets could view the transmission without set-top boxes. Nonetheless, even with digital picture and sound, both digital cable and satellite customers could continue to use their existing analog television sets. The U.S. Congress, however, threw a monkey wrench into the situation by requiring stations to stop broadcasting analog signals as of February 17, 2009.[61] Analog television sets would no longer be able to receive broadcast television without a converter. Analog cable subscribers may also run into problems with local channels and the eventual phase-out of analog cable signals.

As you might expect, there was a great deal of resistance to the idea of making 70 million U.S television sets obsolete. The U.S. law therefore included a provision for $40 coupons to be issued to anyone who requested one for use in defraying the cost of the purchase of a set-top converter.[62] Eventually, as current analog televisions wear out, more households will have digital-compatible television sets, but for now the converter box is a reasonable solution for many.

Television Recording

Before the introduction of the home VCR, television shows had to be watched when they were broadcast. People stayed home to watch their favorite shows. The VCR changed that behavior considerably because it gave people the power to "time shift" broadcast material. They could also save shows to watch them repeatedly.

Television content creators initially reacted negatively to the VCR. Recording television seemed to be a blatant violation of copyright.[63] Universal Studios sued Sony Corp. of America, stating that Sony's device, the Betamax, facilitated copyright infringement.[64] The case worked its way all the way to the U.S. Supreme Court, at which point the Court ruled that making individual copies for time shifting was not a copyright violation but instead was considered fair use of the copyrighted material. The Court ruling also absolved the manufacturers of VCRs from liability arising from any copyright infringement that might be enabled by the machine.[65]

VCRs in the home have been largely replaced with digital video recorders (DVRs). A DVR is easier to program than the notoriously difficult VCR, and it stores far more programming on its internal hard disk than could fit on a videocassette. DVRs also make is easier to fast forward over commercials. However, the precedent allowing time shifting of programming that reaches the home has remained in effect.[66]

---

[61] The law does make an exception for low-power stations. These mostly local broadcasters can continue to transmit analog signals for at least the foreseeable future.

[62] For details on consumer effects of the switch to digital television broadcasts, see http://www.dtvanswers.com/newsroom/dtv_insert.pdf and https://www.dtv2009.gov/

[63] One argument that some people made in favor of time shifting was that commercials would be recorded as well as program content. The advertisers therefore would not be losing any market exposure. However, that reasoning did not take into account viewers with heavy fingers on a remote control's fast forward button. VCRs that theoretically skipped over commercials when recording didn't necessarily work well. They detected the start of a commercial by blank frames. However, not all commercial breaks were preceded by blank frames and not all commercial breaks ended with blank frames. A VCR could get very confused.

[64] The case was Sony Corp. of America v. Universal City Studios, Inc., 464 U.S. 417 (1984).

[65] This decision was echoed in later years when ISPs were held not responsible for the actions of their subscribers when the subscribers were transferring copyrighted material.

[66] There has been a lot of legal action surrounding DVRs. However, it has concerned patent infringement rather than copyright issues.

WHERE WE'VE BEEN

The fields of fine arts and entertainment have been affected significantly by the introduction of technology. As well as using the traditional paper/canvas/paint tools, artists can create artistic content using software. Digital art has become essential to the creation of effective Web pages and to desktop publishing. In addition, there is now recognition that the Web requires the expertise of graphic designers for the creation of effective Web pages.

Photography has moved from a film-based industry to a primarily digital industry. Although there will probably always be film cameras, most cameras sold and used today are digital.

Music has also felt the impact of digital technology. Digital instruments that synthesize sound are used commonly. Fortunately, an expected drop in the need for live musicians has not occurred, even though digital instruments can be computer controlled.

Recorded music, which originally was sold only on physical media, can now be downloaded over the Internet. Some downloading is in violation of copyright, but legal downloads can be purchased. In addition, some artists release their music for free on the Internet to stimulate business for other products.

Filmmaking has been changed by digital technology as well. Computer-generated animation has largely replaced traditional cel-based animation. Recent animated films have included three-dimensional characters that are often difficult to tell from live action. In addition, computer technology has made it easier to integrate animation and live action.

Feature films, once shot solely on film, are increasingly being photographed with digital video cameras. Theaters are converting to digital projection at an ever-increasing rate.

In February 2009 U.S. television broadcasters will be switching from analog to digital signals, as required by law. People who receive television over-the-air and have analog television sets will be unable to receive digital signals and must therefore either purchase a digital television or a set-top converter box.

THINGS TO THINK ABOUT

1. Because it is so easy to modify digital photographs and so difficult to detect that they have been altered, we no longer can trust that any photograph is unaltered. Do you trust news photos that appear in a newspaper or on the Web? Why? What could we do to provide certification that a photo is unaltered?

2. Take a survey of 10 to 20 of your friends or classmates. How many have purchased digital music? How many have downloaded copyrighted music for free? What percentage of each person's digital music collection is legal? What do the results of this small survey suggest to you?

3. Repeat the survey described in the preceding problem using a group of people older than 30 (rather than college students). How do the results differ from the results provided by college students? Does age really make a difference? Do you believe this will change as those who grew up with computers and downloaded music become middle-aged? Why?

4. Is the decision of the U.S. Congress to force television broadcasters to turn off analog broadcasts in favor of digital-only transmissions a good one? Why or why not? (Suggestion: Before you decide, do a bit of research on the differences between analog and digital television transmissions and the capabilities of each.)

5. Using the Internet for your research, assemble a checklist of criteria for good Web page design. Then, go to a major retailer's Web site (for example, amazon.com or walmart.com) and critique it using your checklist.

6. The switch from film to digital theaters is proceeding in the United States at an ever-increasing rate. Research the technology changes a theater must make to be able to show digital films. If possible, find the costs of the equipment. What new demands will this technology place on the projectionists? How will the job of a movie theater projectionist change? How is the change to digital projection similar to the introduction of technology in other fields? How is it different?

7. Assume that you are the CEO of a major record company and are faced with decreasing CD sales, primarily from the switch to downloaded music. Adding to your problems is the growing tendency of artists to bypass a recording company when recording and distributing their music. What strategies might you use to save your business? How will you adapt to the era of downloaded music? What can you do to make your company more attractive to artists? When formulating your answer, keep in mind that your company still has to make money.

8. You have been asked by a graphic design firm to supply them with the latest software for creating digital art. The artists want to be able to paint, to draw, and to lay out Web pages. What software packages will you purchase? Why? How much will it cost for each workstation that you equip?

9. Many DVRs include the ability to transfer programming from the DVR to a computer or a DVD. Is this an activity that should be allowed under the copyright law's "fair use" guidelines? Why or why not?

WHERE TO GO FROM HERE

Battino, David, and Stewart Copeland. *The Art of Digital Music: 56 Visionary Artists and Insiders Reveal Their Creative Secrets.* Milwaukee, WI: Backbeat Books, 2004.

Beaird, Jason. *The Principles of Beautiful Web Design.* Lancaster, CA: SitePoint, 2007.

Benoit, Herve. *Digital Television. Satellite, Cable Terrestrial, IPTV, Mobile TV in the DVB Framework.* 3rd ed. New York: Focal Press, 2008.

Charmasson, Henri J., and John Buchaca. *Patents, Copyrights & Trademarks for Dummies.* 2nd ed. Hoboken, NJ: For Dummies, 2008.

Earnshaw, Rae, and John Vince. *Digital Content Creation.* New York: Springer, 2001.

Gardner, Garth. *Computer Graphics and Animation: History, Careers, Expert Advice.* Washington, DC: Garth Gardner Company, 2002.

Green, Rachel. *Internet Art.* New York: Thames & Hudson, 2004.

Hirsch, Robert. *Seizing the Light: A History of Photography.* Columbus, OH: McGraw-Hill Humanities/Social Sciences/Languages, 1999.

Holmes, Thom. *Electronic and Experimental Music.* 3rd ed. New York: Routledge, 2008.

Jenkins, Mark. *Analog Synthesizers: Understand, Performing, Buying—From the Legacy of Moog to Software Synthesis.* New York: Focal Press, 2007.

Kelby, Scott. *The Digital Photography Book.* Volume 1. Berkeley, CA: Peachpit Press, 2006.

Kelby, Scott. *The Digital Photography Book.* Volume 2. Berkeley, CA: Peachpit Press, 2008.

Mazzioti, Giuseppe. *EU Digital Copyright Law and the End-User.* New York: Springer, 2008.

Noam, Eli M., and Lorenzo Maria Pupillo. *Peer-to-Peer Video: The Economics, Policy, and Culture of Today's New Mass Medium.* New York: Springer, 2008.

Paul, Christiane. *Digital Art.* New York: Thames & Hudson, 2003.

Peitz, Martin, and Patrick Waelbroeck. "An economist's guide to digital music." *http://www.cesifo-group.de/DocCIDL/cesif01_wp1333.pdf*

Price, David A. *The Pixar Touch: The Making of a Company.* New York: Knopf, 2008.

Solomon, Charles. *Enchanted Drawings: The History of Animation.* Revised ed. New York: Random House Value Publishing, 1994.

Sparkman, Donald. *Selling Graphic and Web Design.* 3rd ed. New York: Allworth Press, 2006.

Looking Ahead 12

WHERE WE'RE GOING

In this chapter we will

- Review the issues affecting making future predictions.
- Look at a variety of predictions organizations and people have made concerning the technology of the next 50 years and the ways in which that technology will change us.

INTRODUCTION

In this book we've talked about technological change throughout recent history and its impact on the way we live, work, have fun, and interact with other people. With the gift of 20–20 hindsight, we can see that much of the technology we accept as commonplace today—in particular, the Internet and the Web—took us by surprise. No one was able to predict the way a global network and the linked documents that ride on it would change our lives.

If those who monitor technology missed the Internet and the Web, is it possible for anyone to make good predictions? When we look at predictions of technological change for the next 50 years, should we see them as realistic or fanciful? The first part of this chapter looks at just that question, reviewing the history of predictions and some of what we discussed at the very beginning this book. Then, we will discuss some specific predictions that you can put in your mental time capsule to revisit in years to come.

MAKING GOOD PREDICTIONS: IS IT POSSIBLE?

People have been predicting the future for a very long time. More than 5,000 years ago, for example, the Mayans created a calendar that included a "Great Cycle," ending at the time at which our planet and our sun line up with the center of our galaxy. This event happens only once every 26,000 years; the next occurrence is December 21, 2012.[1] It is unclear whether the Mayans made predictions for the date in 2012 or merely saw it as the end of a natural cycle and the beginning of another, but many others have done so in more recent times.[2] Among the predictions is that the earth's magnetic force will switch poles and there will be major upheaval attendant with that event.[3] An extreme interpretation of the Mayan calendar, however, contends that because the calendar ends in 2012, the world will also come to an end.[4]

The French astrologer Michel de Nostredame (1503–1566), known more commonly as Nostradamus, is also well known for future prophesies. Nostradamus is famous (or infamous, depending on your point of view) for predicting events such as the assassinations of John F. and Robert F. Kennedy, the rise of Hitler, and the end of the world in July 1999.[5] Most of his predictions are vague and couched in language that is open to a great deal of interpretation. Whether he made accurate predictions remains controversial even today.[6]

In Chapter 1 we stated that it was very difficult to make good predictions of the future. Although a few individuals (for example, H. G. Wells and Jules Verne) have been relatively good at predicting future events and technologies, as a whole we humans are notoriously bad when it comes to predicting what will happen with technology. The history of technology is full of predictions that have proven to be wrong, a sampling of which can be found in **Table 12.1**.[7]

Even those who are in the business of watching changes in technology have trouble making predictions. For example, consider the predictions made for 2005 in 1995 by the Battelle Institute staff (**Table 12.2**).[8] If we assume that partial success is a "hit," then six of 10 are hits.

---

[1] For one treatment of why the Mayans may have chosen the winter solstice in 2012 as the end of their Great Cycle, see http://www.levity.com/eschaton/Why2012.html

[2] The Mayans had knowledge of astronomy that wasn't equaled until the 20th century. They had two short calendars, one of 260 days and another of 365 days. In addition, they had the long-range Great Cycle. All of this implies a significant ability to measure the movements of celestial bodies.

[3] The Web site at http://survive2012.com/gery11.php supports the idea that the magnetic poles will reverse themselves in 2012.

[4] See http://www.greatdreams.com/end-world.htm

[5] For opinions that support the accuracy of Nostradamus' predictions, see http://www.didyouknow.cd/nostradamus.htm and http://alynptyltd.tripod.com/

[6] For opinions that debunk Nostradamus' predictions, see http://www.snopes.com/rumors/nostradamus.asp and http://www.allaboutpopularissues.org/Nostradamus-Prophecy.htm. As you read his predictions, consider that when predictions are couched in vague language, it is possible to interpret them in many different ways.

[7] There is an apocryphal story floating around that Bill Gates once stated that no one could possibly need more than 640K main memory in a computer. This is apparently just an urban myth, as Gates himself denies ever having said such a thing.

[8] The Battelle Institute is an international research organization. It manages laboratories for a number of U.S. government agencies, but its focus is global. You can find its Web site at http://www.batelle.org

Table 12.1 A Selection of Famous Technology Predictions That Events Have Proven Wrong[*]

Who Said It (or Where It Appeared)	When It Was Said	What Was Said
U.S. Senator Oliver Smith	1842	After watching Samuel Morse demonstrate the telegraph: "I watched his face closely to see if he was not deranged, and was assured by other Senators as we left the room that they had no confidence in it either."
Editorial in the Boston Post	1865	"Well-informed people know it is impossible to transmit their voices over wires, and even if it were possible, the thing would not have practical value."
William Thomson, Lord Kelvin	1895	"Heavier-than-air flying machines are impossible."
	1897	"Radio has no future."
Thomas Edison	1922	"The radio craze will die out in time."
Thomas Watson, Chairman of IBM	1943	"I think there is a world market for maybe five computers."
Popular Mechanics Magazine	March 1949	"Computers in the future may weigh no more than one and a half tons."
Ken Olson, Chairman of Digital Equipment Corporation (DEC)	1977	"There is no reason anyone would want a computer in their home."

[*] Many of these quotes (and more) can be found at http://www.nsba.org/sbot/toolkit/tnc.html

As you will remember from Chapter 1, true innovative thinking is very rare. Most of the time we improve and refine existing technologies rather than create something truly new. The same seems to be true of future predictions. Predictions of the near future can be fairly accurate because they are based on enhancements of current technologies. The further we go into the future, however, the worse the predictions become because there is time for true innovation to occur. Even predicting just 10 years into the future, the Battelle Institute staff had a success rate of only 60% (a little bit better than chance).

This has never stopped anyone from making predictions. It does mean that we should interpret predictions for what they are: someone's best educated guess at where technology will be going. Fortunately or unfortunately, depending on your point of view, technology has the habit of surprising us with the directions it takes.

Table 12.2 The Top Ten Strategic Technologies by 2005, in Order of Importance, as Assembled by the Battelle Institute Staff in 1995[*]

Prediction	What Actually Happened
1. Human genome matching and genetic-based personal identification and diagnostics will lead to preventive treatment of diseases and cures for specific cancers.	The human genome has been mapped, but we are only now starting to be able to use the information.
2. Super-materials. Computer-based design and manufacturing of new materials at the molecular level will mean new, high-performance materials for use in transportation, computers, energy, and communications.	Although we continue to refine existing materials and are conducting research in the use of nanotechnology, practical uses of the technology hasn't occurred as yet.
3. Compact, long-lasting, and highly portable energy sources, including fuel cells and batteries, will power electronic devices of the future, such as portable computers.	We're nowhere close to this. Laptop computer battery life remains far less than we would like and fuel cells technology is still in its infancy.
4. Digital high-definition television. This important breakthrough for American manufacturers—and major source of revenue—will lead to better advanced computer modeling and imaging.	This prediction is right on target.
5. Electronic miniaturization for personal use. Interactive, wireless data centers of pocket calculator size will provide users with a fax machine, telephone, and computer capable of storing all the volumes in their local library.	Can you say "iPhone"?
6. Cost-effective "smart systems" will integrate power, sensors, and controls. They eventually will control the manufacturing process from beginning to end.	This is happening in many North American factories.
7. Antiaging products that rely on genetic information to slow the aging process will include aging creams that really work.	Don't we wish this had come to pass!
8. Medical treatments will use highly accurate sensors to locate problems, and drug delivery systems will precisely target parts of the body—such as chemotherapy targeted specifically to cancer cells to reduce the side effects of nausea and hair loss.	Although some newer chemotherapy regimens have fewer side effects than those of 10 years ago, we are still far from being able to target cancers without affecting the entire body.
9. Hybrid fuel vehicles. Smart vehicles, equipped to operate on a variety of fuels, will select the appropriate fuel based on driving conditions.	Today's gas/electric hybrid cars do switch between the two types of fuels based on driving conditions.
10. Edutainment. Educational games and computerized simulations will meet the sophisticated tastes of computer-literate students.	This has definitely come to pass, especially for students in grades K–6.

[*] See http://www.battelle.org/SPOTLIGHT/tech_forecast/technologies.aspx for the full range of Battelle predictions.

PREDICTIONS FROM ORGANIZATIONS

Despite our poor track record for making predictions of future technology, we humans can't resist making predictions. The Internet is full of prognostications made by all sorts of people. In this section we look at some of those predictions that have come from organizations with missions that at least in part include keeping track of technological advances and commercial entities (in particular, technology magazines).[9]

The Pew Internet and American Life Project[10] is a nonprofit organization that conducts research on the impact of the Internet on American society and culture. In 2006 it conducted a survey of "Internet leaders, activists, builders and commentators"[11] about how they thought the Internet and the culture around it would evolve by 2020. Over 700 people responded.[12] Rather than give the respondents a completely open-ended forum, the survey first posed six scenarios to which respondents were asked to respond.

A summary of the results can be found in **Table 12.3**. Notice that in no case do more than 70% of the respondents agree with the stated scenario. In fact, in each situation at least 30% seem to be forecasting a more negative future than what is presented in the scenario. The strongest agreement in the survey is perhaps the most frightening: 58% of the respondents see technology resistance becoming violent with an increase in terrorism from resistors.

The Battelle Institute, whose predictions for 2005 appeared earlier in this chapter, has also made 10 technology predictions for a 75-year time frame (**Table 12.4**). Most of these predictions seem quite reasonable. Number 7, however, may come as a surprise to U.S. residents, who have largely abandoned the expansion of nuclear power, but keep in mind that Canada and Europe are far more comfortable with nuclear power. Number 2 is a necessity, although in the foreseeable future biofuels do not appear to be economically feasible for large-scale use.

In 2002 the Battelle Institute released a set of predictions for changes in home-based technologies by 2012 (**Table 12.5**). Notice first that there are no predictions for robotic servants; the favorite prediction from the 1950s has disappeared. Second, notice that many of these predictions are on their way to becoming true. For example, universal remote controls for appliances are available, although they do not operate as smoothly as the remotes we are used to having for electronics.

We are also seeing the emergence of integrated computing, communications, and television products. "Personalized identification" in the form of radiofrequency identification (RFID) chips is a reality, and environmental zones in the home are now available. Voice recognition is also spreading, but it is still relatively primitive,

---

[9] Of course, predictions keep changing. To stay up-to-date with them, see http://www.google.com/alpha/Top/ Society/Future/Predictions/ as a portal to many future-technology Web pages.

[10] http://www.pewinternet.org

[11] Anderson, Janna Quitney, and Lee Rainie. "The Future of the Internet II." Pew Internet and American Life Project. September 24, 2008. http://news.bbc.co.uk/1/shared/bsp/hi/pdfs/22_09_2006pewsummary.pdf

[12] The survey was conducted on the Web and respondents were self-selecting. This means that the sample wasn't random.

Table 12.3 How Respondents Assessed Scenarios for 2020

Exact Prediction Language, Presented in the Order in Which the Scenarios Were Posed in the Survey	Agree	Disagree	Did Not Respond
A global, low-cost network thrives: By 2020, worldwide network interoperability will be perfected, allowing smooth data flow, authentication, and billing; mobile wireless communications will be available to anyone anywhere on the globe at an extremely low cost.	56%	43%	1%
English displaces other languages: In 2020 networked communications have leveled the world into one big political, social, and economic space in which people everywhere can meet and have verbal and visual exchanges regularly, face-to-face, over the Internet. English will be so indispensable in communication that it displaces some languages.	42%	67%	1%
Autonomous technology is a problem: By 2020, intelligent agents and distributed control will cut direct human input so completely out of some key activities such as surveillance, security, and tracking systems that technology beyond our control will generate dangers and dependencies that will not be recognized until it is impossible to reverse them. We will be on a "J-curve" of continued acceleration of change.	42%	54%	4%
Transparency builds a better world, even at the expense of privacy: As sensing, storage, and communication technologies get cheaper and better, individuals' public and private lives will become increasingly "transparent" globally. Everything will be more visible to everyone, with good and bad results. Looking at the big picture—at all of the lives affected on the planet in every way possible—this will make the world a better place by the year 2020. The benefits will outweigh the costs.	46%	49%	5%
Virtual reality is a drain for some: By the year 2020, virtual reality on the Internet will come to allow more productivity from most people in technologically savvy communities than working in the "real world." But the attractive nature of virtual reality worlds will also lead to serious additional problems for many, as we lose people to alternate realities.	56%	39%	5%
The Internet opens worldwide access to success: In the current bestseller, *The World is Flat*, Thomas Friedman writes that the latest world revolution is found in the fact that the power of the Internet makes is possible for *individuals* to collaborate and compete globally. By 2020, this free flow of information will completely blur current national boundaries as they are replaced by city-states, corporation-based culture groupings, and/or other geographically diverse and reconfigured human organizations tied together by global networks.	52%	44%	5%

Exact Prediction Language, Presented in the Order in Which the Scenarios Were Posed in the Survey	Agree	Disagree	Did Not Respond
Some Luddites/Refusenicks will commit terror acts: By 2020, the people left behind (many by their own choice) by accelerating information and communications technologies will form a new cultural group of technology refuseniks who self-segregate "modern" society. Some will live mostly "off the grid" simply to seek peace and a cure for information overload, whereas others will commit acts of terror or violence in protest against technology.	58%	35%	7%

Reprinted with permission of the Pew Internet and American Life Project.

using a small vocabulary for many speakers or requiring a long training period for a single voice with a large vocabulary.[13] It's too soon to judge the other predictions in this group, although the development of fuel cells for personal use now seems further off than 2012.

Not to be outdone by the nonprofits, some commercial enterprises—mostly technology magazines—have made predictions of their own. *Popular Mechanics*, for example, published an article in 2000 that was a follow-up to an article published 50 years previously.[14] In it, they predicted what you see in **Table 12.6**.

PREDICTIONS FROM FUTUROLOGISTS

A futurologist is someone who analyzes and makes predictions of the future. Many futurologists emerge as experts in a specific field and therefore have credibility when making predictions about the direction of that field. Often, once a futurologist has credibility in a particular field, he or she branches out into making predictions in other areas. A futurologist can make a living from predictions by going on the lecture circuit and writing books and magazine articles. The futurologist's credibility is therefore essential to not only his or her reputation but to financial success as well.

There are many people who call themselves futurologists. In this section we look at predictions from some well-known individuals (Ray Kurzweil, Patrick Dixon, Ray Hammond, and Ian Pearson) and where they differ in focus, where their predictions overlap, and as to how far into the future they are willing to make predictions.

---

[13] A number of organizations have switched to voice recognition for their phone services. The problem is that when a voice is too soft, too high, or too low, the system cannot understand. If the telephone menu system doesn't recognize numbers pushed on a telephone keypad as an alternative, there is no way to reach someone via the phone.

[14] http://blog.modernmechanix.com/2006/10/05/miracles-youll-see-in-the-next-fifty-years/

Table 12.4 Battelle Institute Predictions for the Years 2005–2080*

1. Advanced health care, including medical diagnostic and treatment technologies, such as biosensors for both civilian and military applications, vaccines, home medical equipment, and a cure for nicotine addiction

2. Sustainable, renewable energy, such as fuel cells, hydrogen production and storage, solar power, and genetically engineered biofuels

3. Innovative materials, including nanofibers and materials, high-performance and high-intelligent polymers, and biomass products for use in medical, barrier fabrics, and filtration applications, to name a few

4. Mega-data analysis, including modeling simulation, and forecasting of large, complex systems requiring large amounts of data, such as weather forecasting pattern recognition, data visualization voice recognition, and encryption systems

5. Perfecting clean water production and storage technologies that deliver better processes and equipment to the international community

6. Scientific and technical education, including participation and leadership in programs and collaborative research at the local, state, and federal levels

7. Revitalizing nuclear power through innovations in fuel processing, nuclear power, and waste management

8. Managing global climate change using innovative approaches and technologies, including carbon management

9. Continuing to invest in the future through generous support of charitable and civic enterprises on a broader scale

10. Breakthrough welding technology, including computational and computer-based welding and joining techniques that will revolutionize structural stress modeling and fatigue design for industrial and government applications, such as automotive, nuclear, and offshore oil and gas

* These predictions were released in 2004 and can be found at http://www.battelle.org/ SPOTLIGHT/tech_forecast/75.aspx

Ray Kurzweil

Today, Ray Kurzweil is known as a futurologist and a believer in what he calls the "singularity." His first claim to fame, however, came when he invented the first machine to read text out loud. The first version of the Kurzweil Reading Machine was about the size of a refrigerator but today can be embedded in a cell phone. It allows those who cannot see print to pass a scanner over text, which is then translated to speech by the reader and spoken.[15]

---

[15]Ironically—given Kurzweil's views on the future of artificial intelligence—the voice still sounds very mechanical, without inflection.

Table 12.5 Battelle Institute Predictions of the Top Ten Innovations
in Home Comfort and Convenience in 2012

1. Universal control for home appliances. Consumers—even your grandparents—are now comfortable using handheld wireless controls for locking and unlocking their cars, activating automatic garage doors, and accessing their TVs. These devices will become more pervasive because of rapid advancements in microprocessors and wireless communications and will lead to a real "universal control." This is the remote control you'll hate losing the most, as its use will be expanded to include access to computers, lighting, heating, and cooling. It may also be that a laptop or handheld computer will become the universal control for all types of electronics and appliances in the home.

2. Personalized health monitoring and care. Virtual house calls will become a reality as consumers demand quality health monitoring in the comfort, convenience, and privacy of their own homes. Circulatory, heart, or kidney testing will become as easy as home pregnancy tests. Transmit your results to your doctor via the Internet, who can then tell you what to do, or the doctor may elect to send a medical van with professionals to administer more complicated tests or remedies.

3. Home environmental quality. An airtight home is good from an energy savings perspective, but it's not good from an indoor air quality perspective. Consumer demand will lead to improved ventilation along with the energy efficiency that airtight homes bring to homeowners. Indoor air quality will be greatly improved through advanced fans and filters that remove allergens from the air, including outdoor pollens and indoor molds, pet dander, and other particulates.

4. Integration of the TV, telecommunications, and computing. Homes of the future will have access to the most powerful of computers and the most complex of software programs—and often these will be miniaturized to fit into the smallest of electronic devices. This access will come through TVs, cable, or satellite. Computers will be operating throughout the house. Handheld and laptop computers will be as common as telephones. Video telephoning will be done though various methods—TVs, computer screens, mobile telephones—and will become as common as telling time with watches and clocks.

5. Voice recognition and activation. We've come a long way in optical recognition of text, which was very difficult and expensive 30 years ago. It's common today to see scanners, printers, and copiers that have optical readers. The same trend will be true for voice recognition and activation. By 2012, security systems in cars and homes will be activated by your unique voice pattern. You'll be able to speak messages into a computer, which will translate voice patterns into digital text.

6. Personalized energy. Miniaturized fuel cells will eclipse traditional batteries in providing long-lasting power for phones, computers, and electronics. Miniaturized PEM* fuel cells will significantly increase energy efficiency and density of storage. Heating, cooling, and other major appliances could be run from fuel cell power or from the electrical grid. This will bring longer lasting power in greater quantities for electronics and appliances in the home.

7. Environmentally friendly and sustainable materials. Some people hate the "new car" or "new house" smell that comes from carpets and construction materials synthetically made from chemicals. New, more economical materials will be developed using naturally derived fibers, including genetically engineered trees, plants, and crops. These new materials will be more environmentally friendly and acceptable to homeowners. These advancements are part of a long-term trend that favors materials from sustainable sources that can be replaced, unlike the derivatives from oil.

(continues)

Table 12.5 Battelle Institute Predictions of the Top Ten Innovations
in Home Comfort and Convenience in 2012 (continued)

8. Home waste treatment. As Americans become buried in our own waste we will see a backlash against the throwaway society that we have become. From consumer disposables to slash-and-burn fields created from forests, ours is a culture accustomed to throwing away anything and everything. This will change as municipalities continue restrictions on the content and amount of garbage collected. Homes will be required to pretreat solid trash leading to, among other things, a new generation of trash compactors. Homes may also be required to pretreat wastewater.

9. Personalized identification and security. Personalized biochips are just being introduced to the marketplace today. Over the next 10 years, virtually everyone will carry their health and medical records with them, either as a piece of jewelry or as an implanted chip that will be easily read and understood. In addition to vital statistics, bio-IDs will contain your individual DNA master blueprint to help identify people and their medical needs. The ID will also have uses including access to cars, homes, and computers likely in conjunction with voice pattern recognition.

10. Home zone temperature, humidity, and lighting. If you like it meat-locker cool while you're bustling around the kitchen, but your spouse likes it toasty warm while watching TV in the den, new technology is going to help you. Today, HVAC systems heat and cool the entire house from one thermostat location. By 2012 we will see room-by-room heating and cooling for better comfort zones and more energy efficiency. Humidity controls will keep 30% to 50% relative humidity throughout the year. Zone lighting now common in commercial settings will become common in houses with sensors automatically turning lights on and off as you enter and leave the room.

Reprinted with permission from *http://www.battelle.org/SPOTLIGHT/tech_forecast/hightech.aspx.*
* PEM stands for Polymer Electrolyte Member. To find out more about PEM fuel cells, see www.fueleconomy.gov/feg/fcv_pem.shtml

Ray Kurzweil's predictions for the future are based on the impact of the ever-accelerating pace of technological development. He believes that the rate of change increases exponentially rather than linearly. It will eventually reach a point where change is instantaneous; at the same time artificial intelligence will grow beyond human capacity, leading to the "singularity," a point in time at which humans will experience profound changes. The singularity is only a few decades in the future. More specifically, Kurzweil believes that artificial intelligence will have the computing power to think as humans by 2020 and that an artificial intelligence will be able to pass the Turing Test[16] by 2029.[17,18]

Kurzweil's predictions sometimes venture into the realm of science fiction: He sees humans obtaining a type of immortality by downloading their consciousness into machines. In his opinion, this is the logical destination of human evolution. The technology will be available in the second half of this century.

---

[16] The Turing Test specifies that a machine can be considered intelligent when it is impossible for a person interacting with it to determine that the machine is actually a machine. In other words, if a machine can masquerade as a human in an interaction with another human, then it passes the Turing Test.
[17] http://www.newscientist.com/channel/opinion/science-forecasts/dn10620-ray-kurzweil-predicts-the-future.html
[18] The Singularitarians are an organized group of individuals who believe that the singularity is imminent. You can find a copy of the "Singularitarian Principles" at http://yudkowsky.net/sing/principles.ext.html

Table 12.6 Predications Made in 2000 for the Year 2050 by the Staff
of *Popular Mechanics*

Predictions	Status (if applicable)
Pilot-less planes.	
Robot surgeons.	As we noted in Chapter 10, we are already close to this.
New drugs to treat many conditions, including better cancer drugs and the conditions that lead to cancer.	
Growth of replacement tissues outside the body; cloning of replacement body parts.	Stem cell research is working toward this.
Use of animal organs as replacement organs in humans.	This has been tried with pig organs but has yet to achieve long-term success.
Overrides in cars that take control of the car when an accident is imminent.	
Foods that contain substances to improve health, such as chemicals to lower high cholesterol. Very little food will be "bad for you."	
Computers will be a part of household appliances, recording what products you use with them and generating shopping orders automatically.	With the development of RFID tags, this may be possible in the near future.
Glass that changes to admit a preset amount of sunlight.	
Entertainment programming will be delivered by satellite only.	
Television will be replaced by virtual reality and 3D holograms.	Virtual reality has, so far, not lived up to the hype.
People will live to be 150.	

Kurzweil summarizes his position on his Web site as follows:

> An analysis of the history of technology shows that technological change is exponential, contrary to the common-sense "intuitive linear" view. So we won't experience 100 years of progress in the 21st century—it will be more like 20,000 years of progress (at today's rate). The "returns," such as chip speed and cost-effectiveness, also increase exponentially. There's even exponential growth in the rate of exponential growth. Within a few decades, machine intelligence will surpass human intelligence, leading to The Singularity—technological change so rapid and profound it represents a rupture in the fabric of human history. The implications include the merger of biological and nonbiological intelligence, immortal software-based humans, and ultra-high levels of intelligence that expand outward in the universe at the speed of light.[19]

---

[19]For a paper written by Kurzweil explaining his views on change, artificial intelligence, and the singularity see
http://www.kurzweilai.net/articles/art0134.html?printable=1

Patrick Dixon

Patrick Dixon is a physician who has given up the practice of medicine in favor of a career in predicting the future. Although he receives money for lectures, he allows free downloading of videos and entire books that he has written from his Web site.[20] His vision of the technological future includes the following:

- Small computing devices, much like RFID chips, will be embedded in most physical objects, including clothing. The embedded devices will communicate with each other without using the Internet.
- The success of technology innovations in the future is more tied to how people feel about technology than the technology itself. He says: "The future is about emotion—not about technology. It's about how people think and feel in the digital world."[21] This is a major theme throughout Dr. Dixon's lectures.
- Banking and other financial transactions will go wireless and call centers will include video.
- Stem cell research will lead to the growing of replacement organs and tissues.
- Other medical research will lead to life spans of 150 or more years, the ability to grow new teeth, remedies for loss of sight due to old age, and regrowth of hair in places where we want it.
- The disparity between rich and poor nations will destabilize the world.

Dr. Dixon's predictions are heavily focused on technology and business. Most are targeted at the near future, although he rarely gives specific dates for the changes he describes.

Ray Hammond

Ray Hammond is a writer and futurologist well-known in Europe.[22] He publishes many of his ideas in the form of fiction, although his nonfiction has also been widely disseminated. He makes his living from his books and from public appearances. His predictions made in 2005 for the world of 2030 include the following:

- The speed of technological change will accelerate so much that many advances will seem to be happening simultaneously.
- Some time between 2030 and 2040 a technological event will cause human evolution to change fundamentally. (Notice how similar this thought is to Kurzweil's concept of a singularity, a concept that Hammond seems to embrace.)
- Climate change will continue because the effects of greenhouse gasses and other human actions will continue, even if we stopped the activities that impact the environment.
- Artificial intelligence will equal the thought processes of a human and quickly surpass humans.

---

[20] http://www.globalchange.com/
[21] http://www.globalchange.com/technoimpact.htm
[22] http://www.hammond.co.uk/index.html

Ian Pearson

Ian Pearson is a futurologist who works for British Telecommunications.[23] He is the coauthor of a future technology timeline with predictions that include those in **Table 12.7**.[24] In many ways these predictions are more like those made in the 1950s for the year 2000: They mention specific technologies rather than more general trends. Nonetheless, there are distinct similarities with the predictions of Patrick Dixon and Ray Hammond.

OTHER SOURCES OF PREDICTIONS

Futurologists are certainly not the only individuals who make predictions. Sometimes, an organization or person will ask an individual (or group of individuals) to make predictions. For example, in November 2006 *New Scientist Magazine* asked 50 scientists—experts in a wide variety of science and technology disciplines—to make predictions.[25] Predictions of this type are different because the authors are not futurologists by trade and because they are making predictions in a narrowly defined subject area. As examples, consider the following:

- Lewis Wolpert (development biologist): "In the next 50 years, as systems biology and computer models take over, the embryo will become fully 'computable': given a fertilised [sic] egg, with the details of its genome and contents of its cytoplasm, it will be possible to predict the embryo's entire development. From this, new general principles may emerge."[26]
- Steven Weinberg (physicist): "The most important development in physics that I can imagine in the next 50 years would be the discovery of a final theory that dictates all properties of particles and fields."[27]
- Michael Gazzaniga (psychologist): "The next 50 years will focus on the social mind, the fact that humans are social animals and that most of the time our personal mental state is to be thinking about relationships."[28]
- Freeman Dyson (theoretical physicist and mathematician): "The biggest breakthrough in the next 50 years will be the discovery of extraterrestrial life."[29]
- Roger Gosden (physician): "What if we could make 'artificial' eggs and sperm, engineered from the body cells of both parents, which could be combined and implanted into the uterus? Such technology has the potential to conquer infertility, birth defects and genetic disease."[30]

---

[23] See http://www.btinternet.com/~ian.pearson/ and http://www.itwales.com/997789.htm

[24] This timeline has a fascinating and unusual user interface. You can find it at http://www.btplc.com/Innovation/ News/timeline/. It's worth a look just for the fun of interacting with it, if not for viewing the predictions.

[25] You can find these predictions at http://www.newscientist.com/channel/opinion/science-forecasts

[26] http://www.newscientist.com/channel/opinion/science-forecasts/mg19225780.080-lewis-wolpert-forecasts -the-future.html

[27] http://www.newscientist.com/channel/opinion/science-forecasts/mg19225780.077-steven-weinberg-forecasts -the-future.html

[28] http://www.newscientist.com/channel/opinion/science-forecasts/mg19225780.092-michael-gazzaniga -forecasts-the-future.html

[29] http://www.newscientist.com/channel/opinion/science-forecasts/dn10481-freeman-dyson-forecasts-the-future.html

[30] http://www.newscientist.com/channel/opinion/science-forecasts/dn10625-roger-gosden-forecasts-the-future.html

Table 12.7 Selections From the Predictions From the British Technology Timeline 2006–2051

Year	Prediction
2010	• Full voice-interaction with a personal computer • Computer-enhanced dreaming
2012	• Nanotechnology toys • Robots available to replace seeing-eye dogs • Desktop computer as fast as a human brain • Computer-controlled appetite suppressants
2015	• Viewers can choose film roles • Robots that can diagnose their own problems and then repair them • Thought recognition as an input device • Emotion control devices • Mining of helium on the moon
2020	• Ninety percent of humans are computer literate* • Global voting on selected issues • Biochemical storage of solar energy • Holographic TV
2025	• Three-dimensional home printers • Circuits made with bacteria • Direct brain link between human and computer • Factories in space manufacturing commercial products
2035	• Solar power stations • Time travel invented • Space elevator
2037	• Large permanent moon bases
2050	• Telepathic communications • Brain downloads

*The timeline does not state the population for this prediction. Considering that only 82% of the world's population is literate, it would seem that this pertains to those in developed nations.

The question remains, however, as to whether in-depth subject area expertise really makes one a better predictor of the future. As with any prediction for 2050, we'll know when the time comes.

Before we conclude our look at predictions, we should also consider once again the role of science fiction. Most science fiction takes place in the future and usually makes some important assumptions about technology, especially in terms of space travel: A faster-than-light drive is possible. If we put aside the idea that Einstein might have been wrong and the faster-than-light travel can be developed, then we can look to many science fiction novels as predictions of what the authors see for the future of humanity. Because so many science fiction books are set in the far future, the predictions suffer from the same problem that affects individuals and organizations making predictions of future technology: The further ahead in time you predict, the more likely the prediction is to be inaccurate. Nonetheless, science fiction has provided ideas for scientists and technologists and, as we mentioned in Chapter 1, has been the venue for the discussion of technologies that subsequently have been developed.

WHERE WE'VE BEEN

People have been making predictions about the future for a very long time. Some predictions are based on the astronomy of the Mayans who lived 5,000 years ago; the notorious Nostradamus was making predictions during the 16th century. Writers such as H.G. Wells and Jules Verne also made predictions. Today, both organizations and individuals make predictions. However, the accuracy of predictions is often little better than chance, and the question of whether it is worthwhile to make predictions remains.

THINGS TO THINK ABOUT

1. If you were to assemble a time capsule that would be opened in 50 years, what would you put in it? In other words, which of today's technologies would you predict to be of interest and/or importance in 50 years?

2. Consider the prediction scenarios used by the Pew Institute study (**Table 12.3**) with the global predictions made by the Battelle Institute (**Table 12.4**). Which predictions do the two sets have in common? Do you think these are accurate predictions? Why or why not?

3. Does it seem strange to you that someone can make a living predicting the future? Why or why not? Should companies such as British Telecommunications be paying a staff futurologist? Why or why not?

4. Visit the Web sites of three or more futurologists. List the characteristics of a futurologist's Web site that give you confidence in the predictions being made. Justify your reasoning.

5. There are similarities between some of the predictions made by the organizations and individuals you have read about in this chapter. Does being made by more than one source increase the likelihood of a prediction being accurate? Why or why not?

6. Given that we don't have a great track record when it comes to predicting future technology, should people bother to continue making predictions? Why or why not?

7. Now it's your turn: Make 10 predictions for future technology. For each, explain why you think each prediction will come to pass.

WHERE TO GO FROM HERE

Braden, Gregg, Peter Russell, Daniel Pinbeck, Geoff Stray, and John Major Jenkins. *The Mystery of 2012: Predictions, Prophecies & Possibilities.* Louisville, CO: Sounds True, 2007.

Brockman, John. *The Next Fifty Years: Science in the First Half of the Twentieth Century.* New York: Vintage, 2002.

Cristol, Hope. "Futurism is dead: Need proof? Try 40 years of failed forecasts." *Wired,* December 2003. *http://www.wired.com/wired/archive/11.12/view.html*

Kurzweil, Ray. *The Age of Spiritual Machines: When Computers Exceed Human Intelligence.* New York: Penguin, 2000.

Paul, Ryan. "Experts believe the future will be like Sci-Fi movies." *http://arstechnica.com/news.ars/post/20060924–7816.html*

"The World of 2088: Technology." *http://www.washington.edu/alumni/columns/june98/technology.html*

Vass, Thomas E. *Prediction Technology: Identifying Future Market Opportunities and Disruptive Technologies.* 2nd ed. New York: Great American Business & Economics Press, 2007.

Index

The designation "*f*" following a page number refers to illustrations on that page.

Credits

CHAPTER 1

1-1 Used with permission of Electrolux; 1-2 Courtesy of Moller International; 1-3 Courtesy of iRobot Corporation; 1-4 Courtesy of Honda Motor Co., Inc.

CHAPTER 2

2-1 © Lim ChewHow/ShutterStock, Inc.; 2-2 © Kenneth V. Pilon/ShutterStock, Inc.; 2-4 © John Black/ShutterStock, Inc.; 2-5 © Ian M Butterfield/Alamy Images; 2-7, 2-8 Reprint Courtesy of International Business Machines Corporation, copyright © International Business Machines Corporation; 2-9 Courtesy of The U.S. Army; 2.10 Courtesy of the U.S. Navy; 2-11 Reprint Courtesy of International Business Machines Corporation, copyright © International Business Machines Corporation; 2-12 Courtesy of the Hagley Museum and Library; 2-13, 2-14 Reprint Courtesy of International Business Machines Corporation, copyright © International Business Machines Corporation; 2-15 Courtesy of Segway Inc.

CHAPTER 3

3-1 © AP Photos; 3-2 © dba/Landov; 3-3 Courtesy of NASA; 3-4 Courtesy of NASA; 3-5 Courtesy of The Department of Energy; 3-6 Courtesy of International Nuclear Safety; 3-7 © Kim Worrell/ShutterStock, Inc.

CHAPTER 5

5-10 Courtesy of Race Point Group

CHAPTER 6

6-3 Courtesy of John Deere Corporation; 6-4 Courtesy of John Deere Corporation

CHAPTER 7

7-1 © Adrio Communications Ltd/ShutterStock, Inc.

CHAPTER 8

8-1 Courtesy of National Archives; 8-2 Courtesy of Xerox; 8-3 Used with permission of The Tank Museum, UK.; 8-4 © Richard Seaman; 8-5 Courtesy of U.S. Air Force; 8-7 Courtesy of National Archives; 8-8, 8-9 Courtesy of U.S. Air Force; 8-10 Courtesy of Senior Airman Steven R. Doty/U.S. Air Force; 8-11 Courtesy of the U.S. Army, PEO Soldier

CHAPTER 9

9-2 Courtesy of Pennywise Arts

CHAPTER 10

10-1 © Photos.com; 10-2 © Katrina Brown/ShutterStock, Inc.; 10-3 PhotoCreate/Shutterstock, Inc.; 10-4 © Basov Mikhail/ShutterStock, Inc.;

10-5 © beerkoff/ShutterStock, Inc.; 10-6 Courtesy of Intuitive Surgical, Inc.;

10-7 Courtesy of Intuitive Surgical, Inc.; 10-8 © Sebastian Kaulitzki/ShutterStock, Inc.; 10-9 Courtesy of Tescan USA Inc.;

10-10 Courtesy of Louisa Howard, Dartmouth College, Electron Microscope Facility; 10-11 Courtesy of NOAA; 10-13 Courtesy of European Southern Observatory; 10-14 Courtesy of NASA; 10-15 Courtesy of NASA, ESA, M.J. Jee and H. Ford (Johns Hopkins University); 10-17 Courtesy of Fermilab

CHAPTER 11

11-1, 11-2 Courtesy of Wacom Technology Corp.; 11-3 Courtesy of Hasselblad USA

CPSIA information can be obtained at www.ICGtesting.com
Printed in the USA
LVOW110726130513

333472LV00001B/5/P